THE PHYSICAL WORLD OF LATE ANTIQUITY

S. SAMBURSKY

THE PHYSICAL
WORLD OF
LATE ANTIQUITY

Princeton University Press
Princeton, New Jersey

Published by Princeton University Press, 41 William Street,
Princeton, New Jersey 08540

Copyright © 1962 by S. Sambursky
All rights reserved

First Princeton Paperback printing, 1987
LCC
ISBN 0-691-08476-9
ISBN 0-691-02410-3 (pbk.)
Reprinted by arrangement with Routledge & Kegan Paul Ltd., Great
Britain

Clothbound editions of Princeton University Press books are printed
on acid-free paper, and binding materials are chosen for strength
and durability. Paperbacks, while satisfactory for personal collections,
are not usually suitable for library rebinding.

Printed in the United States of America by Princeton University
Press, Princeton, New Jersey

CONTENTS

CONTENTS

PREFACE

THE present book, in a way a continuation of my two earlier ones, describes the development of scientific conceptions and theories in the centuries following Aristotle until the close of antiquity in the sixth century A.D. From the copious literature of that period, and the works of the Aristotelian commentators in particular, I have selected and interpreted texts which are of interest for the comparative history of scientific ideas, with special emphasis on the epistemological foundations of physical theories.

The sequence of the chapters is determined by methodological considerations, but each chapter is largely independent of the others, dealing with one special aspect of the subject. The material contained in some of the chapters is an extension of lectures which I gave in the spring of 1960 at universities in England and the U.S.A.

Relatively little has been published on the history of scientific doctrines in the later centuries of antiquity, with the exception of the relevant chapters in P. Duhem's comprehensive work *Le Système du Monde*, which appeared about fifty years ago. I cherish the somewhat immodest hope that the present book may help to change for the better the attitude of neglect towards later antiquity still shared by many classical scholars and scientists alike. It was a period which retained strong intellectual bonds with the classical past but which, in its mode of thought, was akin to a much later age.

Professor S. Pines of the Hebrew University has read the manuscript and made many helpful comments, Dr. C. Broude has corrected the style, and my wife has given much care to the preparation and typing of the manuscript. To all of them I offer my sincerest thanks.

INTRODUCTION

IN the history of Greek science one has to distinguish between two parallel developments: on the one hand scientific achievements in the technical sense, comprising all the factual discoveries and inventions in mathematics, astronomy and the physical and biological sciences, and on the other hand scientific thought, aiming at the formation of comprehensive theories and the philosophical foundation of a scientific world-picture. The development of science proper, taken in the first sense, gathered momentum in a relatively short period and reached its apex in the third and second centuries B.C. From then on, it slowly declined and, with few exceptions, faded out after the second century A.D. At that time the two greatest scientific achievements of the Greeks— geometry and astronomy—were practically accomplished, and the same can be said of their discoveries in acoustics, optics and mechanics.

Scientific thought, however, continued with uninterrupted vigour from the times of the Milesian philosophers around 550 B.C. until the last Neo-Platonists in the middle of the sixth century A.D. During these eleven hundred years, hypotheses on the creation and structure of the universe were produced, theories on the nature of space, time and matter discussed, scientific concepts formed and analysed from an epistemological and purely logical point of view, and inquiries made into such problems as causality and determinism and the nature of physical action. The main contributors to scientific thinking were not the great scientists themselves, like Euclid, Archimedes, Apollonius and Hipparchus, but rather the founders or representatives of philosophical systems of thought, men like Democritus, Plato, Aristotle, Epicurus and Chrysippus, and in late antiquity the Neo-Platonists. It is worth remembering that of these men only Aristotle, who made the most important contribution to the picture of the cosmos, was also an outstanding scientist, in the field of biology.

Thus in antiquity the two types of scientific activity did not merge into a single stream as they have gradually done in modern

ix

times, since Galileo and Newton. From the seventeenth century on, systematic experimentation and the mathematization of physics were the two main factors which in the physical sciences led to the crystallization of a more or less consistent system of knowledge out of an ever-growing wealth of factual discoveries and the various attempts at a theoretical foundation. To an ever-increasing degree, leading scientists themselves share in the synthesis of the picture of the physical world, having taken the lead from the philosophers in the epistemological investigation of scientific concepts. In ancient Greece the scope of experimental research remained restricted because the Greeks, with very few exceptions, failed to take the decisive step from observation to systematic experimentation. Thus hardly any links were formed between the few branches of science which developed, and they did not expand sufficiently to produce a coherent and inter-dependent system. The body of scientific knowledge in antiquity did not reach the critical mass necessary to induce the great scientists themselves to make an attempt at the construction of a theoretical framework which would unite the results of their own research and that of other branches of science.

In view of these serious shortcomings it is astonishing to what an extent arbitrariness was reduced by the extraordinary flair of the Greek mind for rational speculation in the right direction. Greek scientific thinking elaborated in a qualitative, non-mathematical way two main patterns which became precursors of the basic trends in modern physical thought: continuum theory and atomism. The scientific world-picture of Aristotle, an all-embracing theory loosely related to experience and built on a few theoretical assumptions which were derived in part from earlier conceptions, became dominant in Greek and medieval thought. In fact, it is one of the three major world views in the history of science, being followed after a long interval by that of Newton which has since been replaced by that of relativity and quantum physics. Although the Aristotelian picture of the universe out-lived its ancient rival systems, several of its conceptions and assumptions underwent notable modifications during the more than eight hundred years from Aristotle until the closure of the Academy in Athens by Justinian.

These changes, the subject of the present book, are of great significance for the history of science, for several reasons. In the

first instance they are an eloquent proof of the constant interaction of scientific thought and the slowly but steadily accumu-
lating scientific and technical experience. Most conspicuous of all
is the impact of the progress made in astronomy in the post-
Aristotelian period. Models had to be revised and modified and
efforts were made among the ever-increasing complexity of the
description of planetary motions to invent a unitary explanation
which could match in simplicity that of the discarded Aristotelian model. Further, there was the question which since then
has always occupied the minds of scientists and philosophers alike,
whether these models are only convenient means of illustration,
devices adapted to our needs for an ordered description, or
whether they represent, to a greater or lesser degree, some faithful image of a physical reality corresponding to them.

Several branches of the physical sciences such as mechanics,
optics, pneumatics and hydrostatics, made headway in the
Hellenistic period and often led to conclusions conflicting with
Aristotelian assumptions. These conflicts were resolved in a
variety of ways. Sometimes old concepts, disproved by observational evidence but time-honoured as parts of an accepted philosophical system, continued to coexist with new ones; in other
cases, Aristotle's terminology was extended or given a different
interpretation and thus adapted to the new situation. A gradual
change of scientific language also took place as a consequence of
developments in technology. Some progress was made during the
Hellenistic period in the use of machines and other mechanical
devices, for instance cogged wheels and various types of levers
and pulleys. These technical developments brought about a
greater awareness of the function of mechanisms and created a
mechanistic attitude which was reflected in a corresponding
modification of the meaning of certain Aristotelian concepts.

The two great philosophical systems of later antiquity, Stoi-
cism and Neo-Platonism, had a profound influence on scientific
thought, and were also a direct challenge to the Aristotelian philosophy of science. The Stoics were continuists, like Aristotle, but
their universe was not of a static structure; it was filled with the
dynamic *pneuma* whose tension pervaded space and matter. The
conception of a universal subtle fluid which held together heaven
and earth began to undermine the Aristotelian doctrine of the
fifth element, the celestial aether, and thus threatened one of the

cornerstones of the old philosophy, the dichotomy of heaven and earth. The first onslaught on the aether was made by a Peripatetic in the first century B.C., and it reverberated throughout the following centuries. The idea of the *pneuma* held a central position in the Hellenistic world and it left its marks on such different spheres of human thought as religion—'the wind [*pneuma*] bloweth where it listeth'—and physics where it was the forerunner of the modern concept of a field of force.

Still more striking was the influence of Neo-Platonism on the scientific thought of the last centuries of antiquity; it had a polarizing effect, stimulating as well as paralyzing. Out of the renaissance of Plato there evolved a deep and clear-sighted re-evaluation of the relation of mathematics and science. Never before in antiquity, and not again until modern times, has the function of mathematics as the language of science been recognized so clearly and stated so lucidly. The new analysis of the *Timaeus* did not produce anything creative, but it was of astonishing maturity and insight. The mystical ingredients of Neo-Platonism, on the other hand, its affiliations with astrology and alchemy, the irrational belief in the ultimate unity of the cosmos, had on the whole a retarding and confusing influence on scientific thinking, although even here one can discover exceptions.

This strange mixture of rational and irrational elements, of great analytical power and a tendency to the magical, reflected in a style in which cumbersome and irritating repetitions are intermingled with brilliant formulations, presents a deeply intriguing picture of fascinating attraction. In the very process of its dissolution, the old world once again gathered its strength in a magnificent manifestation of intellectual potency and intuitive understanding. Seen from the viewpoint of the history of ideas it was an attempt to break the spell of Aristotle by a scientific approach which anticipated future events by more than twelve hundred years. The Neo-Platonic conception of mathematics was consciously and eminently anti-Aristotelian, and so was the revolutionary cosmology of the Christian John Philoponus who had grown up as a Neo-Platonist. Future research may shed more light on a period which is hardly known today, an era seen but dimly in the twilight of receding antiquity and the approaching Middle Ages.

I

SPACE AND TIME

---

1. *Absolute and relative space*

IN all periods, philosophers, physicists and psychologists have shared in the analysis of space and time. Some of the greatest minds of Greek antiquity occupied themselves with these central concepts; indeed one of Aristotle's major achievements was his raising of the level of the discussion of these concepts to considerable heights by his own contribution which is such a prominent part of his *Physica*. Aristotle's idea of space—or rather place, as he termed it—stemmed from his concept of continuity. The whole cosmos, finite in extension and bounded by the uppermost sphere of the fixed stars, was a single continuum in which each body at every instant occupied a certain place whose extent Aristotle defined by 'the inner surface of the container', i.e. of the surrounding body. This surface was thus the total sum of all points of contact with the surrounding body which could be either air, water, or earth, or aether, that fifth element of which the celestial bodies were supposed to be composed. Aristotle's conception therefore was that of a coherent entity inseparable from the matter contained in it. The motions of matter are exchanges of place and, in the case of natural motions, are governed by the teleological principle whereby every body strives to reach perfection. This is attained at the 'natural place'—below for the heavy bodies, above for the light ones, and in circular motion for the divine celestial bodies.

There were no explicit definitions of space before Aristotle, but

1

he rejected the notion implied in the doctrines of the atomists and of Plato. For Democritus space was identical with the infinite vacuum which provided the possibility of free motion for the atoms, whereas for Plato, who denied the vacuum, space was the receptacle, the seat of all material happenings.

Simplicius informs us of the conceptual developments of the idea of space from Aristotle until his time in the Corollary on Space contained in his commentary on Aristotle's *Physica* which is the source of most of the passages that will be quoted below. Immediately after Aristotle, two basically different definitions of space were given which formed the prototypes of and were used as patterns for the subsequent theories. They are both surprisingly similar to the two fundamental conceptions of space in modern physics. The first, by Theophrastus, is as follows: 'Perhaps space is not a reality by itself but is defined by position and order of the bodies according to their natures and faculties, as is the case with animals and plants and all non-homogeneous bodies which either have souls or are without souls but have the nature of a structure. For these bodies, too, have a certain order and position of their parts with respect to their whole substance. Thus each, being in its proper place, is said to have its specific order, especially as every part of a body desires and strives to occupy its own place and position' [152].*

This is a clear statement of the purely relational character of space, and is, in essence, the same as that of Leibniz who says, for instance in his third letter to Clarke,[1] 'that I hold space to be something merely relative, as is time; that I hold it to be an order of coexistences, as time is an order of successions. For space denotes, in terms of possibility, an order of things which exist at the same time, considered as existing together, without inquiring into their manner of existing. When many things are seen together, one perceives that order of things among themselves.' Theophrastus, even more outspoken than Leibniz, declares order and relative position of bodies to be the essence of space and thus brings out still more clearly the geometrical aspect of the relation between space and matter which has found its physical elaboration in the general theory of relativity.

* Numbers in brackets after quotations refer to the Index of passages quoted (pp. 181 ff.); superior figures in the text refer to Notes (pp. 176 ff.).

A second statement on the nature of space, quite different from that of Theophrastus, was made by a man who was almost contemporary with him and one of Aristotle's eminent disciples, Strato of Lampsacus (c. 300 B.C.): 'Some make space equal in extension to the cosmical body and declare it, though being void by its own nature, to be always filled with bodies and only theoretically to be considered as existing by itself. This is the opinion of many of the Platonic philosophers and, I believe, Strato of Lampsacus was of the same opinion' [149]. This view of space as an absolute entity is consistent with what we know of Strato's general approach to physical problems. His fragments show him as being influenced to some extent by the atomists in that he postulated the possibility of a vacuum—not a permanent one, as Democritus thought, but a temporary, artificial one, created for instance by man. Strato himself probably proved this by some experiments in pneumatics on which, at a much later date, Hero based his work in this field.

From a historical point of view, we have to regard Strato as the first proponent of a concept of absolute space, in spite of his qualification that this space is always filled with matter. It is essentially equivalent to Newton's definition in his *Principia* that 'absolute space, in its own nature, without relation to anything external, remains always similar and immovable'. That many members of the early Academy, according to Simplicius' authority, were also adherents of the doctrine of an absolute space, is of importance for the understanding of the views of some of the Neo-Platonists. On the whole, the speculations of later antiquity on the nature of space were variations of either Theophrastus' or Strato's ideas, but they were all tinged by the basic conception of Stoic physics and cosmology, the notion of the *pneuma* permeating the cosmos and making it one dynamically interacting organism.

It was the doctrine of the *pneuma* that had transformed the static continuum of Aristotle into a dynamic one. The *pneuma* was in many respects a precursor of the modern field concept; although it was supposed to be corporeal, a very tenuous matter filling the whole universe and interpenetrating all bodies, it fulfilled the functions of a physical field by its tensional qualities and by its capacity to give bodies a coherent structure with well-defined physical properties. The radical view of the Stoics with

3

regard to the interaction of *pneuma* and ponderable matter and, as a consequence of this, the interaction between distant bodies, gave enhanced significance to the relational concept of space. Theophrastus' definition that space 'is defined by the position and order of bodies, according to their natures and faculties', introducing geometrical and group-theoretical characteristics, now merged with the Stoic ideas of an all-pervading force that created that well-ordered continuum called space. The Stoic ideas had an enormous impact on the minds of scientists and philosophers during the last three centuries B.C. They were intensified and modified during the first centuries A.D. by the spread of the fundamental beliefs of the monotheistic religions. *Pneuma* was identified with the divine spirit, and its omnipresence became identical with the omnipresence of God. In conjunction with the relational concept of space this led to the identification of space with God in the doctrines of Hellenistic philosophers such as Philo of Alexandria: 'Place has a threefold notion, first as space filled with a body, secondly, as divine order by which God has totally permeated everything with incorporeal faculties, . . . and its third significance is God Himself who is called place because He embraces the Whole but is not embraced by anything' [105].

A notable example of the combined influence of Stoic and Judaeo-Christian conceptions on the Neo-Platonists is Iamblichus' exposition on the nature of space. Iamblichus lived about A.D. 300 and his writings reveal a very strange mixture of the most abstruse mystical ideas and some very clear and well-formulated scientific remarks that however remained aphoristic in character in the midst of an obscurantist *galimatias*. We shall return to Iamblichus at a later stage, but here some passages on space should be quoted: 'Every body, *qua* body, is in a certain place. Place coexists thus in a natural union with bodies and is never separated from them. . . . Therefore, all those who do not make space akin to cause and drag it down to the concept of boundaries of surfaces or of empty extensions or extensions of whatever kind, are making use of foreign notions and miss the whole purpose of the *Timaeus* which always links nature with creation. One has thus to regard space as depending upon cause, in the same way as bodies were primarily introduced as akin to cause, in the sense in which the *Timaeus* has guided us' [153]. Iamblichus does not mention the Stoics, but that space is akin to cause is an eminently

Stoic idea, even if he tries to link it with Platonic notions. He then attempts to give a definition of space and here his phraseology has a strong biblical flavour: 'Which doctrine will define space completely and in accordance with its essence? That which attributes to space a corporeal force that holds the bodies together and supports them, that raises the falling bodies and gathers together those that are scattered, fills them up and protects them from all sides' [154]. The last phrases are indeed not only strongly reminiscent of, but are literally borrowed from biblical expressions that are used in connection with God. The special Greek verb for 'scatter' used by Iamblichus appears in the New Testament mainly in contexts such as St. John's Gospel 11.52 ('he should gather together in one the children of God that are scattered abroad'). The Neo-Platonists, through their polemics against Christianity, were well versed in the Scriptures and, as this passage proves, consciously or unconsciously adapted some religious associations to their own particular ends.

Two other Neo-Platonists, Syrianus (c. A.D. 400) and Damascius (c. A.D. 525), elaborated on the concept of space as a relational entity, but with special emphasis on the positional aspect. The bodies, so to say, are forced to take up certain relative positions and to arrange themselves in a certain order, and the same applies to the different parts of one body. It was easy to connect this view with Plato's doctrine of the world soul and its creative and harmonizing activities. Simplicius reports on Syrianus' view in the following passage: 'Of those who assumed that space also has form and a power stronger than the bodies, I may mention the great Syrianus, the teacher of Proclus the Lycian. In his tenth treatise on Plato's *Laws* he has written concerning space as follows: 'It is an interval with its own specific distinctions derived from the various orders of the soul and the illumination of the creative forms. It appropriates the various bodies and, with respect to one element, makes itself the proper place of fire . . . and, with respect to another, the proper place of earth . . .' [150]. Thus according to Syrianus natural places derive their existence from the relative arrangement of material objects, this itself being the origin of space. Natural motion and rest in the natural place are caused by the relative positions of all the bodies in the cosmos; they do not have absolute significance, nor are they originated by something inherent in the bodies themselves. The quotation from Syrianus closes with the words: 'Thus neither motion nor rest within the extension are subject to the nature of the bodies nor are they caused by this nature.'

Still more outspoken on this point is Damascius who regards space in its three-dimensional manifold as a kind of matrix which allows for different positions and defines positions in various directions. Space thus becomes a vectorial concept as against the simple concept of length, and is compared with other 'measures' as follows: 'These dimensions, in order not to relapse entirely into the indefinite, are supported by measures: time, as a measure of the activity of motion; quantity, such as number, as a measure of discrete matter; length, such as for instance a cubit, as a measure of continuous matter; and space as a measure of the ramification of position' [151]. This topological conception is elucidated by Damascius in another passage in a slightly different way: 'The motion of a point generates an interval with which space is associated as a measure defining the position of all things in the universe. This measure implies that there exists an extension in three dimensions, i.e. in every direction, and that the whole will be properly arranged everywhere with regard to its own position as well as that of its parts, and that all the elements will have their proper positions each in its own proper place within the universe. Thus if the universe is a sphere, it will always have a centre and a periphery and will be situated in its proper place' [155].

Most of the Neo-Platonic views quoted up to now can, in the last instance, be traced back to Theophrastus' relational conception of space, and the quotations are of interest because they give an idea of the style peculiar to that late period and of the cumulative effect of various influences of philosophical and spiritual systems, either contemporary or of earlier times. We shall now quote passages from John Philoponus (sixth century) who also was brought up in the Neo-Platonic range of ideas but who, in his conception of space, was a strict adherent of Strato's absolute space, filled with bodies, but conceived as an entity in itself. 'Space is not the boundary of the containing body, as one can well conclude from the fact that it has a certain extension in three dimensions, different from the bodies placed into it, incorporeal according to its proper nature and nothing but the empty interval of a body—in fact, space and the void are the same by their

6

nature. . . . How do we explain that bodies exchange their places? If the moving body does not penetrate into another one, and if, further, it is not the surface that moves but the three-dimensional extension, then, if the air is cut through in its place by the moving body, the volumes of the body and of the air exchanging their places must obviously be equal to each other. If now the measure is equal to the object measured, it follows necessarily that if the air measures ten cubic units of length, the amount of space containing it will be the same. It will also hold the same ten cubic units which it had given up before to the moving body. . . . Space is thus cubic, namely of threefold extension, and it is a measure of the objects in space, for it is of the same dimension' [71]. Further on, Philoponus makes the same qualification as Strato with regard to the emptiness of space: 'And I do not maintain that this extension either is or can be empty of every body. This is never the case, for though the void in its proper sense is different from the bodies placed into it, as I said before, space is never devoid of bodies, just as we say that matter differs from form but yet can never be devoid of form' [72].

Finally some excerpts from Simplicius' quotations of Proclus will acquaint us with a unique conception that regards space as a corporeal entity, as a body. Proclus' theory again reflects in a highly interesting way the mental attitude of the later Neo-Platonic thinkers. His surprising conclusion that space must be 'an immovable, indivisible, immaterial body', a definition hardly acceptable to any physicist, is reached by a formal reasoning carried through in strict Aristotelian fashion: 'Space must be either matter or form, or the boundary of the containing body, or the interval between the containing boundaries equal to what is referred to as place' [145]. He rejects the first three alternatives and continues as follows: ' . . . one has to regard the extension between the boundaries of the container as the primary place of every body. But the cosmic extension of the whole universe differs from this particular extension, and thus the latter either exists or does not exist. If it were non-existent, then locomotion would take place from nothing into nothing, for in this case the natural places would not exist, and every motion proceeds according to the nature of something which really exists. . . . If it is existent, it is either incorporeal or it is corporeal. It would be absurd to assume that it is incorporeal. For space must be of the

7

same kind as the object in space, but body and incorporeality are certainly not of the same kind. . . . Thus extension is a body, for space is extension. If it is a body, it is either immovable, or it is moving. If it were somehow moving, it had to undergo loco-motion. Thus space would again be in need of space, and this is impossible, as Theophrastus and Aristotle have shown. . . . If space is immovable, it must be either indivisible by the bodies placed in it, as in the doctrine of the interpenetration of bodies, or divisible, as air and water are divided by bodies moving in them. If, however, space is divisible, the whole will be split and its parts will have a motion relative to the splitting body. It could follow that space is movable, for its parts move. . . . Therefore space must be indivisible. If so, it must be either immaterial or material. But were it material it would not be indivisible. For if material bodies interpenetrate, the result is a division, as is the case when our body is immersed in water. Only immaterial objects cannot be divided by anything, for an immaterial body is impassive, and divisible things are not impassive' [146].

After this reasoning which contains trains of thought sugges-tive of Plato, Aristotle, and the Stoics, Proclus arrives at a result which is eminently Neo-Platonic: 'Summing up all the argu-ments, space is thus an immovable, indivisible, immaterial body. Thus it must obviously be the most immaterial of all bodies. . . . Of all these light is the simplest body (among the elements fire is the most incorporeal, but light is of the essence of fire), and there-fore it is manifest that space is light, the purest among all bodies. Let us now imagine two spheres, one made of light and the other of a manifold of bodies, equal in volume. One sphere is placed around the centre of the world, and the other is immersed in the first sphere. The whole universe will then be seen to be in its place and to rotate in the immovable light. It will not be trans-lated from its place, and in this respect it will resemble space, but each of its parts will be in rotary motion, and in that respect the universe will be less than space' [147].

Space, thus identified by Proclus with light, is elevated to the highest rank of reality accorded to light in Neo-Platonic teach-ings, but his picture of the luminous sphere representing space with all the attributes given to it, and carrying the material sphere of the universe, suggests a specific notion which, as Sim-plicius reminds us, already appears in Porphyry's writings.[2] It is the idea that the soul is clad in a luminous garment called 'the luminous vehicle of the soul', an expression that is also found in several places in Proclus' commentary on the *Timaeus*.[3] In his theory of space Proclus transfers this picture of an immaterial apparel of the soul, made of light, to the world soul, and makes space, the luminous sphere, the vehicle of the material world.

Proclus rejects any comparison of his sphere with the *pneuma* of the Stoics. The *pneuma* was corporeal, and its omnipresence could be explained only by the absurd theory of total mixture. His light, though a body, is immaterial and thus no physical difficulties arise from the immersion of the material world in its continuum: 'The immaterial body does not exert pressure nor is pressure exerted upon it. For any substance on which pressure can be applied must by its nature be able to be acted upon. But light, being impassive, neither divides nor can be divided, and thus there does not follow the absurdity that it interpenetrates totally because of its tenuity. If it cannot be divided, it cannot be chopped into small pieces and in this way penetrate totally into the material world' [148].

One can trace Proclus' idea to Plato (*Laws*, 898), but it is obvious that both his light and the Stoic *pneuma* have a common root, both being derivatives of fire, the purest and noblest element. On the other hand, both can be regarded as precursors of the aether in modern physics, although with a difference. *Pneuma* is to some extent akin to the early conception of the aether, like that of Descartes in his cosmology, a highly tenuous stuff within space. Proclus' immovable sphere of light, however, is space itself; it is the prototype of the aether in the theories of the nineteenth century where it fulfilled the function of an absolute system of reference, i.e. of the primary inertial system which, for instance, gives absolute character to rotary motions.

2. *Absolute and relative time*

The discussions of and the inquiries into the nature of time, an incomparably more difficult concept, had a similar pattern to that of the analysis of space. Again there was a division into camps of relationists and absolutists, and the relational conception again had its variations. Let us for a moment recall, in order to clarify the problem, the parallel division in modern times. Here the absolutist Newton was opposed to the relationist Leibniz whose notions were later supplemented and formulated in the language of physics in the theory of relativity. Newton defined absolute time in his *Principia* as follows: 'Absolute, true and mathematical time, of itself, and from its own nature, flows equally without relation to anything external.' Leibniz, on the other hand, in his correspondence with Clarke, states that 'time, without things, is nothing but a mere ideal possibility', and he claims to have proved 'that instants considered without things, are nothing at all, and that they consist only in the successive order of things'.[4] The theory of relativity finally merged space, the order of co-existent phenomena, and time, the order of successive phenomena, into the spatio-temporal world by an epistemological analysis of the process of the measurement of events in space and time, and by postulating the spatio-temporal distance of two events as the elementary 'interval' of the world.

We know of only one representative of the conception of absolute time in Greek antiquity, and this was the same Strato (*c.* 300 B.C.) who also defined (in a qualified way) the absolute character of space. Simplicius, in his Corollary on Time which forms a chapter of his commentary on the *Physica*, quotes Strato's definition of time. This is not only an excellent paraphrase of its absolute conception, it also gives a clear representation of time as an independent variable, with other parameters depending on it in different degrees: 'We say that we are abroad or are sailing or serving in the army and fighting in a war for a long or a short time, and similarly we say that we sit or lie still and do nothing for a long or a short time. Where the quantity involved in these is large, a long time has passed, and where the quantity involved is small, a short time has passed. For in all these cases time is a quantity. This is why one says that time passes slowly or quickly, depending on how large a quantity is involved in each of these happenings. We say time passes quickly when the quantity from the start of the events to the end is small, but the events that have occurred are many; and we say on the contrary that time passes slowly when a large quantity has passed with little happening. Thus there is no speed or slowness in rest, for rest is completely equal to the quantity of time passed, and not long in a small quantity or short in a large one. That is the reason why we talk of longer or shorter time but not of quicker or slower time. Action and motion are quicker and slower, but the quantity in which action takes place is neither quicker nor slower but is more or less, and this is time. Day and night, a month and a year are not time nor parts of time, but they are light and darkness and the revolutions of the moon and the sun. Time, however, is a quantity in which all these are contained' [160].

The last sentences of this passage show clearly that Strato, like Newton, distinguishes between the absolute flow of 'the river of time' and the events occurring in time which are like leaves or branches carried along with the stream. But of much greater significance is the earlier part of the passage where Strato speaks of the rate of action in a given time. We must remember that the Greeks never reached the stage of a precise scientific definition and mathematical formulation of even the simplest physical quantities such as velocity or acceleration. Further, they developed only the most rudimentary concepts of functional thinking and especially did not visualize changing entities as variables depending on time. In his very concise sentences Strato comes nearest to this when he considers the ratio of actions to the time they take. He actually transcribes into ordinary language a

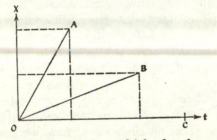

modern graphical representation which, for the case of velocity, would appear as in the diagram. If the vertical axis represents distance and the horizontal time, then OA represents the ratio of a comparatively large distance to a short time, i.e. a large velocity, and OB the ratio of a small distance to a large time, i.e. a small velocity. Strato chooses actions as his dependent variable and speaks of the number of events that occur in a given time. These actions or motions, he says, are quicker or slower depending, as we would say, on the steepness of the lines OA or OB; time itself, however, is only a quantity, namely a certain abscissa on the horizontal axis. Strato considers 'no action', or rest, to be a special case; rest, he says, is exactly equal to the quantity of time passed, which is a transcription of the fact represented in the graph by a horizontal line, e.g. OC along the time-axis. One can hardly overrate this achievement of Strato who was the first to express clearly—as far as this is possible in non-mathematical language—that rest is motion along the axis of time.

Strato's eminently physical conception of rest as a function of time[5] had a certain impact on subsequent scientific thought. Two hundred and fifty years after him, similar ideas were expressed by the Peripatetics. Boethus (c. 50 B.C.) also recognized that motion and rest have one thing in common, namely, the change of the independent variable of time, whereas they differ in that only in the case of motion the time-dependent entity changes with time. Simplicius gives the following account of Boethus' reasoning: 'It is not correct to describe rest in a place as "place". For Boethus regards states of rest not as negations of the respective motions, but for him rest and motion are relationships between time and the parameter to which motion or rest is ascribed. He says: "It seems that the nature of time is an eternal flow and change into different states, and therefore time accompanies every motion and rest. Thus motion behaves equally with regard to time and place, but rest behaves differently with respect to them. For in the latter case the time is never the same, but the place is, and therefore the very relation to both time and place differs in the cases of rest and of motion, the same being true also for other parameters. When there is the same relation of matter to both time and size, we call it motion, and when it is contrary with regard to size, it is called rest; the same can be said of qualitative changes. From all these, it is manifest that rest does not mean the same size, the same form or the same place, but it means a relation of everyone of these with regard to time"' [140].

It is no mere coincidence that Boethus, like Strato, regarded time as an absolute entity, as Simplicius relates on another occasion[6]. This conception of time as something independent of events, eventually facilitated its visualization as a 'quantity', a co-ordinate to whose changes can be related such changing events as actions and motions. Strato and Boethus arrived here at the threshold of the conception of a physical quantity in the modern sense, a concept built up of combinations of entities of different dimensions such as the ratio of distance and time. They might easily have crossed this threshold, had the Greeks invented symbolic algebraic notation and its application to objects of the physical world.

The relational concept of time has its origin in Aristotle's writings, although his elaborate exposition on time in the fourth book of the *Physica* is not too clear on this point. However, as we shall see, it has given rise to very important developments of thought in several directions. The pivotal sentence on which Aristotle's discussion of time turns is his definition: 'Time is just this—number of motion in respect of "before" and "after" ' [14]. He thus takes the 'now' as the point of reference to which the flux of time is referred, but he gets into difficulties by whittling down this 'now' to a mathematical point, and thus reducing it to a mere limit between past and future. What happens between two different 'nows' is actually the essence of time. However—and this makes Aristotle a relationist—he clearly recognizes that time and motion are interrelated. In the same way that space exists only in so far as there are bodies which occupy a certain place, time exists only in so far as there are bodies which, at different 'nows', are in different places or in different states. Time is not a quantity in itself, as Strato saw it at a later date, but it cannot be separated from the motion of physical objects. 'Just as the moving body and its motion involve each other mutually, so too do the number of the moving body and the number of its motion. For the number of motion is time, while the "now" corresponds to the moving body and is like the unit of number' [15].

Aristotle was thus quite consistent in his relational approach both to space and time. The essential difficulty of his point-like 'now'—particularly awkward for a relationist because it deprives the present of all reality—was removed by the Stoic doctrine of time. In strict conformity with their dynamic conception of continua, the present of the Stoics *qua* limit of time is not sharp but forms a fringe that covers the immediate past and future. Chrysippus (third century B.C.), a contemporary of Archimedes and the main pillar of the older Stoa, arrived at this conception of the present by a limiting process of infinite convergence that consists in an approach to the mathematical 'now' both from the direction of the past and from the future. The present is thus given by an infinite sequence of nested time intervals shrinking towards the 'now', whereby the lower boundaries of each interval are points in the past and the upper ones points in the future: 'Chrysippus states most clearly that no time is entirely present. For the division of continua goes on indefinitely, and by this distinction time, too, is infinitely divisible; thus no time is strictly present but is defined only loosely' [42]. From this 'loose' definition the present emerges as a shrinking duration with only indistinct boundaries. The physical significance of such a duration is that it represents an event-like structure, it is an elementary event, and thus macroscopic time, in the strict relational sense, can be conceived as being composed of the succession of such events.[7]

Damascius, as we have seen before, specified the relational conception of space by regarding it as a measure defining the position of all things in the vectorial sense. In the same sense he amplified on the relational conception of time: 'Just as the parts of separate things do not overlap because of space, the occurrence of the Trojan war does not become mixed up with that of the Peloponnesian war because of time, nor in a man's life does the state of a new-born become mixed up with that of a youth' [158]. And similarly Proclus declares that 'time signifies the progression in an order and the well-ordered descent' [117].

3. Time and change

The philosophers of late antiquity, and the Neo-Platonists in particular, were however mainly occupied with other aspects of the problem of time. There was for instance Aristotle's statement that 'not only do we measure motion by time but also time by motion'. Is this reciprocity something peculiar to time and motion, or does it hold also for other standards of measure and objects measured by them, in other words—is it generally true that the measure and the measured exchange roles? We can for instance measure a certain distance by a yardstick and then use this distance to gauge a rod of unknown length. Philoponus makes some suggestive comments on this: 'Later, Aristotle says that not only time measures motion but that reciprocally time is measured by motion. For we say that much movement has taken place if we measure this movement by time and if much time has passed, and again we say that much time has passed if we measure this time by motion and if much motion has taken place. Similarly is the jug measured by the wine and *vice versa*. For we determine the size of a jug when measuring it by a certain amount of wine, and again we determine an amount of wine by measuring it in a jug. We also say that a given corn-measure measures a certain amount of corn and equally we determine the corn-measure by measuring it with a certain amount of corn. But actually it is only the measure that is determining the corn or the wine, and it is itself not determined by them. For even granted that the wine measures the jug and the corn measures its measure—this can only be done after they have been measured before by another measure, because the measure is the essential means of measurement. . . . But even though a given corn-measure, by way of a measured amount of corn, is thus measured by another corn-measure, which is the same as what was said before that the corn-measure is determined by a certain amount of corn and that a certain amount of corn is measured by the corn-measure—this does not hold in the case of time. For motion is not first measured by time and after that motion measures time, but they are devised for each other like relative concepts. Time and motion are defined by each other like the father who derives his existence *qua* father from his son and the son his existence *qua* son from the father, both simultaneously' [80].

In the elaborate and repetitive style of his time Philoponus emphasizes the unique position of time and motion among the family of standard measure and measured objects. They are intertwined and cannot be disentangled because they mutually presuppose each other. Time is measured by a moving mechanism and, to be sure, by a mechanism whose motion is periodic. The prototype of all periodic mechanisms, of all clocks, is the revolutions of the outermost celestial sphere and the more intricate periodic motions of the planets. As Aristotle remarks, this is 'why time is thought to be the movement of the sphere, namely because the other movements are measured by this, and time itself is measured by this movement'.[8] The most celebrated literary exposition on this subject in antiquity was of course Plato's *Timaeus* (37–38) which was echoed in Plotinus' chapter on Time and Eternity and in Proclus' commentary. Proclus tells us that the sun which marks the most conspicuous periodic interval of time was called 'the time of time'.

The strict periodicity of the heavenly standard of time and the affiliated periodicity of the seasons were no doubt the main reason for the belief in the eternity of the world, and it became a scientific dogma in Greek antiquity since Aristotle and was only challenged by the monotheistic dogma in the early centuries A.D. It is interesting to note how deep-rooted became this belief in the eternity of the world in pagan antiquity, induced as it was by the daily and yearly cycles. There is no sign of any idea of a running-down of the universe, of any precursor of the concept of entropy, in the whole scientific thought of Greek antiquity before the intervention of Christianity in the later writings of Philoponus. Even the Stoic conception of *ekpyrosis*, the evolution of the universe towards a state of the prevalence of the hot element, which they attempted to prove by geological and meteorological evidence, was part of a cyclic theory on a vast scale. After *ekpyrosis* will have reached its climax, the cold and wet elements will again become dominant and the whole thermic cycle will repeat itself, even to the extent of the return of the identical. Periodicity on a small and on a large scale, and the return of the identical, must have been a welcome counterbalance to the conception of an eternally lasting universe, existing without beginning and without end. The curious asymmetry of the Aristotelian cosmos—finiteness with regard to space and infinity with regard to time—lent additional weight to the analysis of time, the infinite extension of becoming embedded in the infinite being of eternity.

This brings us to the relation of time and change, a problem much debated in later antiquity with greater intensity and ingenuity than the question of absolute and relational time. Time and change is a subject that throws into strong relief the interference of psychology with the physical notion of time. Time is a much more complicated concept than space because it is invariably connected with consciousness, with the apprehension of an eternal and continuous flux which is independent of our five senses. We could rather imagine space empty of all matter than time flowing on without events, without change, because the very flow of time is bound up for us with the experience of the flow in our consciousness, identified by the ancient Greeks with the soul.

Aristotle touches upon the relation of time and the soul only in passing, and his answers are only aphoristic in comparison with the questions he raises, but he says in another context that even when we are not affected through the body, we associate 'any movement that takes place in the mind' with some lapse of time. It is characteristic of Neo-Platonic thought that this 'inner movement' of the mind or the soul was seen by Plotinus as the real essence of time. His famous sentence that 'time is the life of the soul as it moves from one state of life to another' [110] omits the physical aspect and is a purely psychological definition. Today, Plotinus' criticism of Aristotle's definition of time may seem somewhat unfair to those who are inclined to assume that the 'movement of the soul' cannot be dissociated from the—periodic and aperiodic—motions of molecules which are of the same nature as the external motions that Aristotle referred to in his analysis of time.

Leaving aside the problem of time and consciousness, the overriding question pertaining to time and change still remains: is time real? Does it exist or not? Aristotle has expressed it very well at the beginning of his inquiry, and it was this that bothered the Neo-Platonists most: 'The following considerations might make one suspect that time either does not exist at all or scarcely and in an obscure way. Some part of it is past and no longer exists, and the other is future and does not yet exist; yet time—whether infinite or any time we can consider—is made up of these. One could hardly conceive of something made up of non-existent things having any share in reality' [13]. These words of Aristotle were a challenge to all those who lived under the spell of Plato's celebrated metaphor on past, present and future being the forms of time when imitating eternity. The feeling that there is something unreal about time, and the desire to discover in it some status of reality involved the Neo-Platonists in endless discussions which lasted from Plotinus to Simplicius, a period of well over three hundred years.

That these philosophical speculations also concern the physicist is obvious to any student of physics and of the theory of relativity in particular. In recent years McTaggart's well-known essay on the unreality of time has been quoted for instance by K. Goedel in connection with a relativistic model of the universe which he constructed and which allows for the unusual, though theoretically valid possibility of a return into the past.[9] What degree of reality can be attributed to a fundamental physical concept when it can lead to consequences where past and future become mixed up? It is therefore not without interest even to the scientist of today to look into some of the Neo-Platonic discourses on time. Of special significance are the ideas of Damascius who was a contemporary and friend of Simplicius and is quoted by him at some length. Damascius' considerations have a twofold purpose—to assign to time some sort of reality and to avoid the pitfall of an infinite regress inherent in the concept of time. This latter which is also one of the objects of McTaggart's analysis consists in the necessity of defining a second time-series in order to explain the first series of futurity, presentness and pastness of which each event is a member and which is one of the basic characteristics of change in time.

One of the passages where Damascius discusses the ontology of time ends with a typical Platonic figure of speech: 'Even if time and motion are in a continuous flux, they are not unreal but their being consists in becoming. Becoming however is not simply non-being but it is existing in always different parts. . . . And it is evident that time coexists everywhere with motion and is coherent with change, in that everything in time has its existence in becoming or, to put it otherwise, that time makes becoming circle round being' [157]. In spite of change being the essence of time Damascius believes in the conservation of the identity of a thing, because of the continuous nature of that change. Simplicius' account continues as follows: 'When Damascius says that "time itself is at the root of changelessness of things that depart by themselves from their present being, and thus time contains rather an element of rest than of motion"—he apparently meant that there is continuity also in becoming, because of the similarity of time and eternity, for as eternity is at the root of rest in being, so time is at the root of rest in becoming. . . . But these words of his do not disturb me as much as what he often said to me when he was still alive, without however convincing me—that the whole of time has a simultaneous existence in reality' [159].

Simplicius is not able to accept his friend's conception of the totality of time in its continuous flux as a single undivided reality. Damascius clearly recognized the difficulties involved in this notion which are caused by our inability to put ourselves outside time because we are forced to relate events to the present as the zero-point of our temporal co-ordinate system. Damascius gives other instances of this tendency of the soul to introduce specific structures into entities that in reality form a unity, and then he continues as follows: 'Exactly like that, I believe, the soul attempts to bring to a standstill the river of becoming by the stationariness of the forms inherent in the soul and, by delineating three parts of time that are fixed in relation to the present, apprehends them as separate manifolds. For the soul, having its place in the middle between the reality of becoming and that of being, attempts to comprehend each of them in accordance with its own nature; it decomposes the objects of being on a level inferior to that of being, but more natural to the soul, and it unites the objects of becoming on a level superior to that of becoming but more cognizable to the soul. Thus it recognizes day and month and year by gathering each of them together into one form and delineating segments of the whole flowing time' [161].

Because of our failure to grasp the totality of the river of time as a single unfolding reality, change of time is bound up for us with the succession of futurity, presentness and pastness, perceived from the shifting zero-point of the Now, and this leads to the difficulty of infinite regress, described by Damascius with brilliant lucidity: 'If the Now perishes, it either perishes in itself or in another Now, because everything perishing perishes in time as everything coming into being does so in time. But evidently this theory of time presupposes time. Indeed, by this reasoning one could prove that it is the motion of a motion. And generally, when we try to take the measure of measures we go on infinitely taking one standard measure as a gauge for another standard measure, and one number for another' [162]. Damascius, unlike McTaggart, tries to arrive at a solution by cutting the Gordian knot: 'There is no necessity for time to perish in time and the Now in another Now, nor is it possible that several times should exist simultaneously. For the reality of the Now is to be seen in the flux of time in relation to a presupposed rest of whatever kind. Should somebody ask if time, having its being in becoming and moving itself, is not in need of another time for measuring and arranging its parts so that they do not get mixed up, the answer is that time moves so that in accompanying motion it becomes a measure of motion. For the standard measure, too, exists separately in addition to the measured object and preserves the specific individuality of the standard and is not in need of another measure' [163].

It is difficult to understand the meaning of the 'presupposed rest of whatever kind' by which Damascius wants to bring to a halt the infinite regress. As a disciple of Plato he believed in the mediating role of the soul between the eternity of being and the ephemeral becoming and on this, as we saw, he based the faculty of the soul to recognize macroscopic stretches of time within the incessant change of the Now. The last sentence of the passage seems to indicate that Damascius exchanged the philosophical plane for the purely physical and laid down by definition an ultimate standard of time. However, the main crux, so clearly shown by Damascius, still remains unresolved—that we can look at events in time only in relation to the zero-point of a fleeting Now.

II

MATTER

---

1. *The mechanistic theory: conceptual developments*

THERE is hardly a chapter in the history of physical thought that has greater fascination for the physicist of today than the earlier ideas on the constitution of matter and the nature of change in matter. Quantum theory gave us the first real glimpse of the essence of matter; now, some sixty years later, we feel that we are only at the beginning of a long and exciting journey. From our present vantage point we realize the extent to which attempts at an analysis of matter are bound up with profound changes in our conception of physical reality and the nature of scientific explanation. Everyone who believes that there is a certain inner logic in the history of scientific thought and that definite trains of thought repeatedly force themselves upon the scientist on different levels of knowledge and insight, will be amply rewarded by a study of the ancient Greek ideas of matter.

Looking back on Greek antiquity as a whole, one can distinguish between three different approaches—the mechanistic, the qualitative and the mathematical. The first was that of the pre-Socratic atomists Leucippus and Democritus, the second was Aristotle's doctrine, and the third was Plato's geometrical theory. All three occupied the minds of post-Aristotelian thinkers to a varying extent and at quite different periods of later antiquity. In this age, the mechanistic theory found its most prominent representative in Epicurus (300 B.C.) and still flourished in the first century B.C. with Lucretius as its major proponent; however, it later sank rapidly into obscurity, although Democritus and Epicurus were still referred to quite frequently by the Aristotelian commentators.

The qualitative approach of Aristotle was accepted and cultivated by the Stoics and became the dominant doctrine, as can easily be seen in the writings of Plutarch, Galen, Alexander of Aphrodisias, and others.

Plato's geometrical theory of matter was pushed into the background by Aristotle and practically fell into oblivion until the rise of Neo-Platonism. Then, after over seven hundred years, it was revived, put on a firmer conceptual foundation and stood in high esteem during the fifth and sixth centuries A.D. Nevertheless, it was Aristotle who held the field for almost the whole period of the Middle Ages.

The ideas of Epicurus keep within the conceptual framework of the mechanistic theory of atoms as laid down by Leucippus and Democritus, but in his writings there are some deviations from the original doctrine, one of them involving general principles. This is of great interest and importance because it is one of the rare cases in Greek science where conceptual development can be studied in a well-defined theory, worked out in detail right from the beginning. The point on which Epicurus differed from Democritus is well known—it is the question of the sizes and shapes of the atoms. Democritus had assumed that atoms could be of every conceivable size, and he did not exclude 'very large' atoms, or even atoms of 'cosmic dimensions', which implies that he believed that atoms could be visible. This assumption that there is no upper limit to the size of an atom was a direct consequence of Democritus' assumption that the number of atomic shapes is infinite, and of the basic presupposition of atomism that all changes are discontinuous. For if one postulates an infinite variety of shapes that do not form a continuous set, but in which each shape differs from the others by a finite step and is thus clearly distinguishable from every other shape, one is obviously forced to admit also an infinity of sizes. This was precisely what was most emphatically rejected by Epicurus who saw in the assumption of macroscopic atoms a violation of a basic principle of atomism: 'In each shape the atoms are absolutely infinite in number, but their differences of shape, though incomprehensibly large, are not absolutely infinite, unless one is prepared also to keep enlarging their sizes *ad infinitum*' [45].[11]

What were the reasons of Leucippus and Democritus for their postulation of an infinity of atomic shapes? 'They say that the number of shapes of the atoms is infinite, for there is no reason why anything should be of one shape rather than of another' [141]. It was thus the principle of sufficient reason which was applied here in its negative form, as in other instances that will be discussed in Chapter IV. In cases where a supposed symmetry or equality prevails, the conclusion seems possible that nature lacks sufficient reason to prefer certain alternatives to the exclusion of others, and as one geometrical figure is as good as another, there could be no limitation of the shapes of the atoms in the opinion of the early atomists.

The opposing point of view of Epicurus bears on the fundamental dichotomy of the atomic theory in general—the clear-cut division between the realm of physical occurrences not directly accessible to our senses, and the range of macroscopic bodies. His refutation of Democritus' opinion is formulated in his letter to Herodotus as follows: 'Again we should not suppose that every size exists among the atoms in order that perceptible things may not contradict us' [47]. The sort of contradiction Epicurus has in mind is hinted at in a passage which comes shortly before the one quoted: 'Moreover, we must suppose that atoms do not possess any of the qualities belonging to perceptible things . . ., for every quality changes' [46]. The essence of *phainomena*, of bodies and processes accessible to the senses, is that they have definite qualities which can undergo changes. However, the main attribute of atoms which is also the cause of their being indivisible and stable, is their impassibility, and a visible atom would mean a body not susceptible to change and stripped of every possible quality. Such a body would indeed contradict experience and would be something 'which is never observed to occur, nor can we conceive how its occurrences should be possible, namely that an atom should become visible' [48].

It should be noted that the impassibility of the atoms was regarded by the atomists as the ultimate reason for the persistence of the macroscopic properties of substances and for the permanency of natural phenomena. The indivisibility or the absolute rigidity or hardness of every atom was an immediate result of this impassibility. It was Epicurus who shifted the emphasis from indivisibility to impassibility, and the arguments of later opponents of atomism were directed not so much against the concept of an extended but indivisible entity as against the dialectic picture of a macroscopic body endowed with qualities but composed of particles devoid of any quality. This was formulated very clearly by Sextus Empiricus as the antithesis of the whole to its parts: 'Epicurus maintained that the part is other than the whole, as the atom is other than the compound, since the former is devoid of quality whereas the compound has qualities, being either white or black or, generally, coloured, and either hot or cold or possessed of some other quality' [135].

Epicurus thus regarded his epistemological considerations, by which the size of an atom was restricted to the range inaccessible to our senses, as stronger than the metaphysical principle of sufficient reason invoked by his predecessors. His conclusion was therefore that there can exist only a finite number of shapes among the atoms, all of them beyond the range of visibility. But Epicurus did not stop at this point and, once the principle of sufficient reason was overthrown, he imposed further restrictions on the shapes of atoms by which certain types of shapes were ruled out: 'The shapes of atoms are of an incomprehensible but not infinite number. For there are neither hook-shaped ones nor trident-shaped ones nor ring-shaped ones because these shapes are easily breakable, and atoms are impassive, unbreakable' [115]. This assumption is in clear contradiction to that of Democritus who said that 'some of the atoms are rough, some hook-shaped, some concave and some convex and there exist other, innumerable shapes' [208].

Epicurus established a selection rule for atomic shapes, differentiating between 'allowed' and 'forbidden' shapes and excluding the 'easily breakable' ones. This is clearly bound up with a more mechanistic view, with a definite step towards a purely mechanical conception of the atom. Impassibility had been conceived in the first instance as an abstract geometrical property by which atomic shape was invariably fixed once and for all. Democritus had already used the term 'solidity' or 'rigidity' synonymously with impassibility in the sense of geometrical invariability, which was not meant to be a mechanical property opposed to softness or deformability. Epicurus however came to regard impassibility as something akin to mechanical firmness and solidity and used the expression 'impassive' in combination with the term 'unbreakable'.

It does not seem excessive to presume that Epicurus drew analogies between atoms and macroscopic bodies as regards the relation between shape and mechanical firmness. 'Primary' qualities such as hardness have always been regarded by the atomists as inherent, 'objective' properties, like geometrical structure, and it is moreover known that from the purely geometrical standpoint Epicurus tried to explain extension and magnitude of atoms by analogy with macroscopic particles. As the number of atomic shapes was supposed to be limited in Epicurus' theory, it was not too bold a step to postulate further restrictions based on mechanical considerations.

It is hardly possible to draw definite conclusions from the one extant passage, but it is tempting to speculate as to the possible theoretical background of Epicurus' 'selection rule' for atomic shapes. Besides evidently being guided by mechanical analogies he may have based his notions on a generalization of the fundamental postulate of Greek atomism, that 'the atom does not share in the void'. The presupposition of complete separation of the substance of atoms and the void was necessary for the explanation of atomic stability. Similarly, Epicurus may have 'forbidden' atomic shapes exhibiting too great a ramification of structure, such as protrusions or indentations or annular forms. He may possibly have regarded such shapes as 'easily breakable' and not stable, because they increase the surface of contact between the atom and the empty space surrounding it. In view of Epicurus' rather unmathematical mind it seems doubtful whether in this connection he considered the singular position of the spherical shape with its extremal property of minimum surface at given volume. Considerations of this kind, however, played a significant part in the later Hellenistic period, as we shall presently see.

Another example of a conceptual development in the Greek theory of atoms concerns a typical case of the later reaffirmation of a hypothesis by a discovery made in another field. The early Greek atomists, in order to explain the most conspicuous qualities of the fiery element as well as of the mind, had made two postulates with regard to the atoms of fire and those of the soul. They supposed them to be spheres and thus to have great smoothness and mobility, and further assumed that they were of small size, which makes plausible the tenuous substance of fire and the even more tenuous property of the soul.

It was only through further progress made in geometry about a hundred years after Epicurus that the conceptual significance of the spherical atom could also be given a more concise expression. This progress is connected with the discovery by Zenodorus of the isoperimetric theorems, probably during the second or first century B.C. Theon of Alexandria gives an account of it in his commentary on Ptolemy's *Syntaxis*, and it is also briefly mentioned by Simplicius. Zenodorus proved that the area of a circle is greater than that of any regular polygon of equal perimeter, and that the volume of a sphere is greater than that of any other solid body which has an equal surface area. About four hundred years later, Pappus restated these theorems and showed that the sphere is greater than any of the five regular solids having the same surface, and also greater than either a cone or a cylinder of equal surface. Extant sources do not reveal whether after Zenodorus the isoperimetric theorems were applied by the Epicurean School to the problem of spherical atoms.

Philoponus, from the perspective of the sixth century A.D., characteristically disregarded the chronology of this mathematical development that had taken place long after Democritus and Epicurus. In retrospect, he attributed to Democritus isoperimetrical considerations as his motive for ascribing spherical shapes and smallness to the atoms of fire and the soul: 'It is worth while to inquire into the reason for Democritus' statement that spherical atoms consist of the smallest parts and are thus easily mobile. It is evident that the spherical shape is easily mobile for it can be shown that the sphere touches a plane only in one point. But why are the spherical atoms composed of the smallest parts, which is supposed to be another reason for their mobility? This does not seem to make any sense at all, for pyramid-like and hook-like atoms can also be composed of very small parts. However, the explanation is as follows: in geometry it is shown that among the rectilinear shapes of equal perimeter those with more angles have a larger area. . . . Thus the circle will have the greatest area, because the more angles a shape has the nearer it comes to being without angles, i.e. to being a circle. The same law holds for solids, and thus the sphere has the largest content among all the rectilinear solids of equal surface. But if this is true, i.e. if among the isoperimetric shapes those with more angles have a larger content, the reverse will also hold, namely that of all shapes of the same content those with more angles will have the smallest perimeter. The spheres will therefore possess the smallest surface of all. Democritus thus rightly assumed that of all atoms equal in volume the spheres are the smallest and for this reason able to penetrate everywhere . . .' [91]. The assumptions of Democritus and Epicurus thus seemed vindicated in the eyes of Philoponus because shape and (relative) size of the spherical atoms are coupled by definite physico-mathematical relations.

Finally we will take a look at the problem of the mixture of two liquids, a typical and often discussed chapter of the theory of matter, on which Epicurus' opinion differed significantly from that of his predecessor. We are told of this by Alexander of Aphrodisias, the orthodox Aristotelian: 'According to Democritus there exists no real mixture at all, but what seems to be a mixture is a juxtaposition of bodies that preserve each their specific nature which they possessed before mingling. They seem to be mixed because of the failure of our senses to perceive the separate bodies lying side by side for reason of their smallness. . . . Epicurus refused to follow Democritus in this conception. He too belonged to those who regarded a mixture as a juxtaposition of bodies, but he declared that these bodies are not preserved in the process of division but dissolve into elements and atoms. Each of the bodies that consist either of wine or water or honey or of something else is somehow composed of these elements and atoms. From the quality of these bodies of which the components consisted the mixture results by synthesis. Thus it is not water and wine that mingle but, so to say, water-producing and wine-producing elements, and the mixture results from a kind of destruction and generation. For the dissolution into the elements of each component is destruction, and the synthesis of the components is generation' [5]. We see here Democritus holding a view deriving from the assumption that the quality of a mixture must be the result of the mixture of the qualities of its components, whereas Epicurus took the radical stand of an uncompromising atomist. In his theory each component is broken up into its ultimate units, impassive and devoid of quality, and each of them is surrounded by units of the other component, yet the atoms of both components add up and combine to form the quality-possessing mixture.

Here we are confronted with a special case of the general question—how the macroscopic, physical properties of matter are built up from the basic, geometrical data of shape, order and position of the atom. This question again belongs to the class of problems, often put as paradoxes, how and at what stage quantity turns into quality. In principle the atomists assumed that the cumulative effect of the atoms leads to the transition from the impassive and unchangeable individual entity to the compound exhibiting quality and change. 'Not every single spherical atom is fire, but the summation [*soreia*] of all of them' [88]. The term *soreia* suggests the famous dialectical argument *soreites* often mentioned in Greek literature, to the effect that a state reached by the gradual addition or substitution of small units cannot be sharply defined. How many grains add up to a heap? How many hairs must be lost before baldness is reached? Curiously enough there seems to be no reference in the extant sources to the problem of atoms—how many atoms together add up to a body endowed with qualities? This state must of course be reached well below the threshold of visibility, because all visible things exhibit change and quality.

In this connection one must point at an essential weakness of Greek atomism that was not overcome in Hellenistic times. Democritus as well as Epicurus presupposed only the existence of atoms and the vacuum and omitted all continuous forces from their theory of matter. It is true that in addition to the instantaneous forces of impact resulting from collisions, Democritus also assumed interatomic forces of a sort, namely the interlocking of atoms of a suitable shape, e.g. hooked atoms. But he regarded these bonds as only temporary, being broken up and formed again by the constant whirling around of the elementary particles. Epicurus, on the other hand, in his selection rule, rejected precisely those shapes that allow for the formation of bonds. Aristotle in his criticism of the atomists recognized this weakness of Greek atomism and asked how the atomists explained the difference between water and ice.[12] There are obviously the same atoms in the same arrangement both in water and in ice, and the difference between the liquid and the solid state thus remains unexplained.

We see here a characteristic difference between ancient and modern atomism. The modern atomists have retained in their theories an essentially continuous concept, that of force. Interatomic and nuclear forces are an integral part of physical reality for the quantum physicist; they are continuous entities describing physical fields around particles which are singularities in these fields. Physics today cannot do without the dualistic approach of atomism and continuum concept. One of the factors which might have prevented the Greek atomists from contemplating continuous forces was their basically mechanistic approach. Continuous forces could easily be identified with spiritual entities like gods or demons, and the interference of such entities with the picture of the material world had to be strictly avoided.

2. *Plato's geometrical theory and Aristotle's objections*

The first outline of a mathematical theory of matter was given by Plato in the *Timaeus*.

Plato does not devote a single word to Democritus and his theory. This silence was possibly an expression of contempt for the mechanistic and materialistic approach of the atomists who wanted to explain spiritual reality by mere kinematic and mechanical hypotheses. Furthermore Plato was influenced by the Pythagorean School, that pre-Socratic train of thought which, in contrast to the mechanistic views of the Milesians and of Empedocles and Anaxagoras, made number the basis of all physical events.

However, Plato's theory has one basic presupposition in common with that of Democritus: the hypothesis that the phenomena of the macroscopic world are rooted in some discrete invisible elements whose agglomerations and interactions cause all physical occurrences. In this respect Plato's theory of matter is an atomic theory and it is significant that in antiquity Plato and Democritus were often mentioned together when the explanation of matter was discussed. Still, there are two important differences: Plato denied the vacuum, and he built his atomic elements from a much more restricted range than that of the immensely great number of shapes postulated by the atomists.

Another question is how far the relevant passage in the *Timaeus* can be taken as a scientific document. Plato likes allegory, and his poetical vein is apt to interfere with his scientific train of thought. The question of the interpretation of the *Timaeus* is largely a question of attitude towards Plato's scientific intuition. If one takes him as seriously in this respect as one takes him as a philosopher one must look beyond his poetic language and take it for granted that he meant quite definite and basic things in what he said. This was certainly how the Neo-Platonists saw it when they revived his theory some seven hundred years later.

The outlines of a mathematical, or rather geometrical, theory of matter are given in the sections 53–58 of the *Timaeus*. In view of the brevity of Plato's exposition it should perhaps be called a blueprint of a theory which he may have put down in the hope that someone more expert than he would work it out in greater detail. He did not claim that his suggestion was the only possible or correct one, but he was convinced, in true Pythagorean spirit, that certain symmetries prevailed in the structure of matter and that therefore the proper approach to the explanation of matter was the geometrical. The symmetry must be found in the realm of tri-dimensional geometry, because the extension of matter is tri-dimensional. It was a completely legitimate approach for Plato to choose the four perfect 'Platonic' bodies for this purpose and to identify them with the four elements. He knew already that there are five perfect bodies but he omitted the one for which he had no use in his construction of the elements of matter—a procedure for which no modern theoretical physicist can possibly blame him especially when remembering Plato's words that he would welcome as a friend anyone who offered a solution better than his.

In this well-known scheme, Plato associates the tetrahedron with fire, the octahedron with air, the icosahedron with water, and the cube with earth. Thus one of the principles of the mechanists was upheld in that shape was the determining factor for the physical behaviour of the elements, but for Plato this was only the starting-point for more far-reaching assumptions. One important aim he wanted to achieve with his theory was the explanation of material change, for instance such a basic phenomenon as the transition from the liquid to the gaseous state (the Greeks did not distinguish between water vapour and air), or the

30

reverse process. Therefore he had to look for some features common to all or to some of the four elements, and to see if they provided a key for the explanation of change from one element to the other. The tetrahedron, octahedron and icosahedron are all bounded by equilateral triangles, and this immediately allows for the establishment of relations between them and thus between the elements which they represent. On the other hand the cube is bounded by squares that cannot be resolved into equilateral triangles by further division. Plato therefore concluded that no

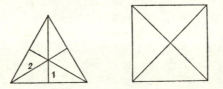

transition can occur from earth to one of the other three elements fire, air and water. Even so, Plato did not choose the equilateral triangle and the square as the ultimate structural elements but subdivided them into rectangular triangles. Equilateral triangles can be constructed from rectangular ones whose smallest and longest sides have the relation 1 : 2, and squares are composed of triangles which are both rectangular and isosceles. From these two types of rectangular triangles, the planes bounding the three first bodies as well as those of the cube can be composed.

One advantage of the decomposition into small structural elements was that sets of equilateral triangles or of squares of different and increasing sizes could be constructed, which again enabled the construction of sets of elementary bodies of different sizes. Plato could thus differentiate, for instance, between different sorts of fire, including light, or different sorts of air according to the size of the composing elementary bodies. But evidently within each set the tetrahedron is the smallest body because it is composed of the smallest number of triangles. This immediately yields two main characteristics of fire, its great mobility (smallness of size) and its penetrability (acuteness of the solid angle). Here two properties of fire for which Democritus had to stipulate two separate conditions (smallness and sphericity) were reduced to a single root.

The dissolving property of the most active of all the elements is thus explained by the great acuteness of the tetrahedral shape which can easily pierce and dissolve the other bodies. The geometrical counterpart of this dissolution is the decomposition into the elementary triangles out of which other bodies reconstitute themselves according to fixed numerical relations. When a cube is dissolved it can only be rebuilt into a cube, as we have seen. But when an icosahedron is pierced by a tetrahedron, the twenty equilateral triangles can reconstitute themselves into the boundaries of two octahedra (16 triangles) and one tetrahedron (4 triangles). One atom of water is thus transformed into two atoms of air and one of fire. In the reverse process two parts of fire can for instance combine into one of air, or five parts of air into two of water.

There is no need for us to discuss here all the details of Plato's theory, as we shall return to it when we come to describe its revival by the Neo-Platonists. What is important at this stage is to realize that Plato's mathematical approach to the problem of matter achieved, though in a rudimentary and primitive form, what the mechanistic theory did not: it connected features of material properties and of material change with certain geometrical and numerical relations. In a way, this was the beginning of mathematical physics in its most elementary form in that certain elements of the material world were described by mathematical quantities which in turn allowed for the establishment of equations or relations that could be interpreted as data and processes within the physical world.

Aristotle, as he is wont to do, gives a critical review of the theories of his predecessors, including Plato, as an introduction to his own qualitative theory. He refers to Plato's geometrical doctrine several times, in the *De caelo* as well as in the *De generatione et corruptione*, and some passages in the *Metaphysica* have an indirect bearing on the subject. Seen from a wider perspective —as by the Neo-Platonists—the main issue between Aristotle and Plato derives from their different views of the role of mathematics in the physical sciences or, more generally, of the relation of mathematics and science. Aristotle occasionally had recourse to mathematics in order to explain certain physical facts, for instance in his discussion of motion. But on the whole, mathematics, and geometry in particular, was to him nothing more than perceptible things seen in abstraction from their perceptible qualities. There are many illuminating passages in his writings where this is stated quite clearly. The mathematician investigates abstractions, he says in the *Metaphysica*: 'He first strips off all the sensible qualities such as weight and lightness, hardness and its contrary, and also heat and cold and all other sensible contrarieties, leaving only quantity and continuity . . . and does not study them in any other respect' [39]. One of Aristotle's illustrations of mathematical abstraction, the 'snub-nose', has become famous: 'Of things defined, i.e. falling under the category of "what", some are like "snub" and some like "concave". The difference is that "snub" is a combination of matter and form, because what is "snub" is a concave nose, whereas concavity is independent of perceptible matter' [38].

Aristotle was aware of the fact that mathematics is applied to different branches of science, to astronomy, optics, mechanics and harmony, and that 'it is the business of the empirical observers to know the facts, of the mathematicians to know the reasoned fact' [7]. Physical sciences when applying mathematics study matter and form, but they do so by treating physical realities in respect of certain attributes abstracted from their physical essence. However, it never occurred to Aristotle that mathematical elements, for instance geometrical shapes, could be used as symbols to describe physical realities. This was precisely what Plato did in the *Timaeus* and what is at the bottom of Aristotle's objections to his theory, not only where they are of a principal nature but also where they refer to technical details.

It is significant that when Aristotle compares Plato's theory with that of Democritus he gives preference to the latter while of course rejecting both: 'To resolve bodies into planes and no further does not make sense, as we have also said elsewhere. Hence there is more to be said for the view that there are indivisible bodies, though this also involves considerable absurdity' [29]. By its very nature, every atomic theory has to presuppose certain least elements that are unchangeable, and here Democritus' assumption of minimal solid particles seemed more reasonable to Aristotle than Plato's idea of choosing two-dimensional shapes as the elements of a tri-dimensional continuum. Aristotle did not grasp that Plato's elements were geometrical shapes and not material, despite Plato's statement that fire and earth, water and air are bodies, and that every body possesses solidity, and every solid is bounded by planes, which implies that these planes are to be taken in a purely geometrical sense. Plato never thought of attributing to his triangles weight or any other material property. However, a considerable part of Aristotle's polemics is based on the erroneous imputation that Plato regarded the planes as material. He points out for instance in the *De caelo* that solids could also be built up from planes in a different way, namely by piling them up instead of putting them side by side, and the construction of the elementary bodies out of planes would thus not be unambiguous. This and similar objections show that Aristotle missed the essential point of Plato's theory which is that certain regularities of matter and material change can be derived from geometrical relations between the surfaces of bodies that symbolically represent the four elements.

3. *The qualitative theory*

The discussion of a number of further objections raised by Aristotle will be postponed until we come to the Neo-Platonists who entered into a detailed argument with his criticism. Here we are concerned with the basic attitude of the realist and experimentalist against the idealist whose ultimate principle he believes to be wrong. In a passage which rather detracts from Plato, Aristotle confronts his 'dialectical' method of inquiry with the 'physical' method of Democritus: 'Those whom devotion to long discussions has rendered unobservant of facts, are too ready to dogmatize on the basis of a few observations' [30]. Aristotle's strong bias against Plato can be understood more easily in the light of his own principle of scientific explanation which he defines thus: 'Perceptible things require perceptible principles, eternal things eternal principles, perishable things perishable principles; and, in general, every subject-matter principles which are homogeneous with itself' [26]. Neither Democritus nor Plato conformed to this rule. The atomists had sinned against it in that they had discarded explanation by perceptible qualities and explained matter by elements corporeal and not perceptible, by atoms devoid of any quality; but even greater was the sin of Plato who discarded explanation of matter by corporeal elements and explained bodies by surfaces or, ultimately, by two-dimensional geometry.

Aristotle in his theory of matter assiduously followed his principle of explanation. Starting from the four elements fire, air, water and earth he attempted to explain the changes occurring in matter by the assumption that these four elements are nothing other than combinations of two of the four elementary qualities hot, cold, moist and dry. By forming the four possible combinations, Aristotle attains four elements, and by further assuming that every quality can be replaced by its opposite through the action of another body he explains in principle the transition of one element into another and material change in general. The ultimate elements of the perceptible physical world are thus not corporeal, but they are perceptible as the sensible qualities of heat, cold, humidity and dryness.

Aristotle's conceptual system was generally accepted, it outlived the others and persisted, with modifications, until the beginning of modern times. The main reason for this longevity was its success in biology and medicine where it served to put current ideas on a systematic foundation. The medical theories of the four humours and subsequent developments in ancient and medieval physiology were related, in one way or another, to the qualitative doctrine of matter. Aristotle himself applied his system to biology and in the *De partibus animalium* he explains the significance of the four qualities for biological processes.[13] One can easily recognize their share in the parts and organs of the animal and human body when one studies tissues, bones and blood. Aristotle analyses in particular the meaning of heat in this connection and points out that everything that grows and therefore takes in food, must transform the food in some way. Hereby heat fulfils a vital function, which is why animals and plants must have some natural source of heat in themselves. Later, however, he says that plants make use of the heat of the earth which prepares their food in the necessary form.

Theophrastus, in the footsteps of Aristotle's researches on animals, wrote his books on plants and for the first time established botany as a systematic scientific subject. He took over the Aristotelian notions and applied them to the world of plants. In his work *De causis plantarum* he discussed among other topics the growth of plants as related to climate, soil, water and other environmental factors. The four elements are obviously involved in these processes (in which fire is represented by the sun) but what about the function of the four qualities? In contrast to the decidedly perceptible character of animal heat, it is difficult to regard warm and cold as sensible qualities in plants. 'As regards growth and early sprouting of plants, one has to see the cause in the influence of air and sun and in the specific nature of each of the plants which differ in their humidity, dryness, degree of density and similar factors, as well as in heat and cold. For these, too, are part of their nature. But while humidity and dryness are more or less accessible to the senses, heat and cold, which do not involve perception but reason, lead to differences of opinion and controversies as does everything that is subject to the judgement of reason. It is worth while to define these concepts, especially as many things are referred to these principles. Indeed, one must judge all such concepts by their factual consequences, for from the facts we come to conclusions about the nature of elementary forces' [239].

We have here an interesting development in scientific thought which has also become familiar to us in modern science. Concepts and modes of explanation become more involved and abstract when, in the course of scientific progress, they are applied to an expanding field of knowledge. The concept of inertia for instance was originally associated with motion at a constant velocity in a straight line, and has been retained in general relativity in the more abstract sense of motion along a geodetic line. Also many of the classical concepts of Newtonian mechanics which were taken over by quantum mechanics have lost some of their concrete significance. From Theophrastus' words we learn that the concept of heat, when applied to plants, has ceased to be a 'perceptible principle' in the Aristotelian sense and has become a subject of controversy among experts. Later on Theophrastus polemizes against an earlier botanist, the Pythagorean Menestor who lived in the second half of the fifth century B.C. and was influenced by Empedocles in his ideas on the action of heat and cold. Empedocles had assumed that animals which have a preponderance of one quality in them are driven into regions where the opposite quality prevails. The general idea was, as Aristotle remarks in the *De partibus animalium*, the compensation of one quality by the opposite one. Theophrastus puts it thus: 'Things perish through the same quality as a result of hypertrophy and are saved by the opposite quality through the production of a certain well-balanced mixture' [240]. He then continues: 'The same view

36

was held by Menestor with regard to animals as well as to plants. The warmest plants, in his opinion, are those which grow in water, like rushes, reeds or galingale, and which also are not frozen by winter storms' [241].

Theophrastus rejects this criterion of warmth as well as another which connects fertility with heat as an inner quality of plants. He insists that only the affinity of inner quality and quality of environment is conducive to generation, nourishment and preservation. On the other hand, he puts forward another criterion that introduces a new and original point of view: 'One has also to ascribe the quality of heat to those faculties which produce heat, fermentation and humours in bodies when the plants have been consumed as food, or which create the sensation of heat when tasted or touched. This needs no confirmation by scientific evidence, as it is proved by medical practice and by sensation' [242]. The first part of this passage, which relates the heat content of a plant to its metabolic effects, makes heat a chemical concept; thus the concept of a 'hot plant' was given by Theophrastus a 'perceptible' and measurable significance, although in an involved and sophisticated way which was most probably not what Aristotle had had in mind. When Theophrastus associated heat with chemical processes it was certainly more than an aphoristic remark. We know from his writings, from his discourse *On Fire*, for example, that he was well aware of the various aspects of heat and its different effects, depending on the nature of the thermic process. The following passage may serve as an illustration: 'Fire or the quality of heat thus admits of great differences in itself as well as within, or in conjunction with, other bodies. Probably, this fact also explains the problem of why the intestines can dissolve coins while boiling water which is still hotter cannot do it' [244].

The Stoics took over the qualitative theory of matter and thus assured the continuity of its tradition, notwithstanding some modifications which they introduced in accordance with their physical doctrine. Their all-pervading *pneuma* was a mixture of fire and air, and its most conspicuous property, tension, was determined by the active and elastic properties of its two components. Fire and air were therefore regarded by the Stoics as active elements, in contrast to the passive ones water and earth. In parallel with this classification they distinguished between active qualities—hot and cold—and passive ones—dry and moist—and related each of these qualities to one of the elements. The Stoics conceived of everything as a body capable of acting or being acted upon and they had no use for the Aristotelian distinction between corporeal elements and incorporeal qualities. Every element was thus characterized by, and indeed almost identified with, its predominant quality: fire was hot, air cold, water moist and earth dry. The *pneuma*, too, whose total mixture with matter conferred upon the bodies their physical properties, sometimes figured as a substance and sometimes as a fifth quality. The Stoics, of course, had no use for the Aristotelian aether either, because their doctrine of *ekpyrosis* assured a periodic transition of the cosmos as a whole from its ordered state into a state of thermal chaos, out of which order evolved again, leading to a return of the identical.

All these modifications were regarded as serious deviations from the Aristotelian conception and led to endless polemics in the late Hellenistic period. Historically seen, they do not alter the main fact of the qualitative character of the Stoic theory of matter in which the basic approach was still that the qualities are the fundamental principles of the material world, and that their mixture is the mechanism by which every material change in the realm of the inorganic as well as the organic is produced. The medical theories dating back to Hippocrates, long before Aristotle, and up to Galen in the second century A.D., were in strict conformity with this conception and thus contributed most effectively to the firm establishment of the qualitative theory of matter. Besides Plutarch's essay on *The Principle of Cold*, there is ample proof of this in the writings of Galen, and among these his work *On Mixtures* deals exclusively and at great length with the application of the qualitative theory to biological and medical problems of the human body.

This work, which consists of three books, shows clearly how solidly Galen's descriptive medicine and his physiology were based on the doctrine of the four qualities, already time-honoured in his period, as stated in the opening of the book: 'That the bodies of living beings are a mixture of hot, cold, dry and moist, and that their share in the mixture is not equal, has been sufficiently shown by the foremost philosophers and medical men of older times' [50]. What makes *On Mixtures* such interesting reading is that it offers a most revealing example of the historical development of a fundamental set of concepts. In the course of a long tradition, the four qualities as applied to an increasingly complex set of biological phenomena became rather involved and abstract notions which sometimes were only remotely connected with their original significance. On the other hand, a strict empiricist like Galen tried hard to uphold the links with the sensory origins of those concepts and even attempted to construct physical standards by which they could be converted into measurable, reproducible quantities.

The following passage shows how these concepts were relativized and adapted to the organic world: 'If one says that bodies are mixtures of the hot and the cold, the dry and the moist, "bodies" in this context is to be understood in the absolute sense of the word, namely elements—air, fire, water and earth. If on the other hand one talks of an animal or a plant as being hot or cold or dry or moist, one must understand these words differently. For a living being cannot exist in the absolutely hot state like fire, nor in the absolutely moist state like water. And similarly, there exists in this case no extreme cold or dryness, but rather are the denotations formed according to the component which predominates in the mixture. For instance, we call a substance moist when it contains a greater share of moisture, and dry when the share of dryness is greater. And the same with hot and cold' [51].

The next passage introduces the concept of a 'normal' type or mixture on which Galen puts great stress in his book: 'As there are many classes of animals and also many individuals, the same body can be hot, cold, dry or moist in many ways. When one draws comparisons with an arbitrary case, it is obvious that the same thing can be called by opposite terms, e.g. that Dion is dryer than Theon or Memnon, but more moist than Ariston and Glaucon. But when comparison is made with the normal case of a genus or a species this can easily perplex and confuse the layman. For one and the same man can at the same time be moist and hot or dry and cold, namely the latter when compared with the normal type of man, and the former when compared with an animal or a plant or any other substance' [52].

Galen emphasizes the complexity of the concept of a 'normal mixture' when applied to biology that, on the psycho-somatic level, leads to the definition of the four temperaments. 'We say that all animals and plants have an optimal and normal mixture within their own class, not simply when there prevails exact equality of the opposites but when there exists a certain proportion with regard to their faculties. It is similar to the sense in which we say of justice that it establishes equality not by weight and measure but according to propriety and merit. Equality of mixture exists in all well-tempered animals and plants and is not defined by the weights of the components of the mixture but by a measure that fits the nature of the animal or the plant' [53].

The large measure of flexibility thus introduced into the conception of the qualities made it possible to adapt a terminology of long standing to the needs of biology and medicine which indeed was done by Galen with admirable skill. Curiously in contrast to this sophisticated procedure is Galen's somewhat naïve attempt at a standardization of the four qualities: 'In every class, and in all structures generally, the mean is generated by the mixture of the extremes, and it is thus the extremes from which one has to derive the conceptual content as well as the diagnosis of the mean. We can exactly determine the mean by conceiving the distance between the hottest perceptible quality such as fire or fiercely boiling water, and the coldest which is known to us, such as ice or snow. We will thus arrive conceptually at the symmetrical value which is equally distant from both extremes. We can moreover construct it somehow by mixing equal masses of ice and boiling water. The mixture of both will be equally distant from both the extreme of burning and that of mortifying cold. Further, it will not be difficult, once the mixture is apprehended by our senses, to define the mean of every other substance with regard to the opposites hot and cold, and to measure everything by comparing it with a recorded standard. Similarly when soaking earth or ashes or some other very dry matter in an equal quantity of water, one will construct a substance which is the mean of the opposites dry and moist. Again, it will then not be difficult to identify such a body by sight or by touch and to keep it on record and use it as a standard and criterion for the diagnosis of substances that have a deficiency or an excess of moisture or of dryness' [54].

Galen is well aware of the fact that the consistency of such a paste defining the mean of dry and moist depends on its temperature. He therefore continues: 'Obviously, the standard substance must possess normal warmth. For if the norm of moist and dry is too hot or too cold it may present a false picture and appear sometimes more moist than the norm and sometimes drier. If it is hotter than the norm its appearance will be more dissolved and fluid than the moist. On the other hand if the norm is colder than it should be it will become stiff and motionless and will appear rigid when touched and thus create the false impression of dryness. When the standard body presents the mean of hot and cold exactly as that of moist and dry it will appear neither rigid nor soft to the touch' [55]. What is again so typical of later Greek antiquity is the striking contrast between the advanced conception of a standard and the primitivity of the suggested experimental procedure. However one must remember that the idea of a standard already had a long history in aesthetics dating back to the time when classical Greek art was at its peak. Galen himself reminds us of this association of concepts when he says in the same context: 'So do painters and sculptors paint and mould objects representing the most beautiful type of every species, such as the best-shaped man or horse or bull or lion, by looking for the norm in each class. And somehow the statue of Polycleitus is praised and called a standard, because all of its parts have perfect proportions in respect to each other' [50]. Symmetry and proportion are central concepts where the scientific mind of the Greeks met and matched in full harmony with their aesthetic feeling and ethical attitude.

There can be no doubt about the overwhelming influence of Galen in establishing the preponderance of the qualitative theory of matter which held good for nearly 1500 years. In this respect, despite many eclectic features in his theoretical views, he was a firm follower of Aristotle and the Stoics and adhered to the principle of qualitative explanation, as for example in the following passage: 'The actual diagnosis of the hot, dry and moist is easily accessible and well known to everyone. The distinction is by touch which teaches that fire is hot and ice cold. If there are people who can derive the concept of hot and cold from some other source they should tell us. If as a criterion of perceptible things they offer something different and more important than perception they propose an impracticable art or rather, to tell the truth, something stupid. . . . If they look for abstract proof in matters of perception they could as well examine whether snow is white as it appears to all men, or whether one should regard it as non-white, as Anaxagoras declared' [57].[14]

Galen's rather primitive attempt at a quantitative definition of a mixture of qualities was destined to remain almost unique in this field. Once again the problem of the quantitative composition of mixtures was taken up much later by Philoponus in quite another context. He posed a question which in this form had never been raised before in antiquity. Assuming that the physical properties of a substance all result from a given mixture of the elementary qualities, Philoponus asks how one can explain that

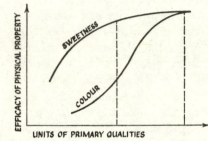

in some cases one of the properties may visibly change while the others apparently remain unchanged. Formulating this question was in itself a novelty, but the answer he gives implies for the first time the discussion of the functional dependence of one set of variable quantities on another and the clear recognition of the course of a function, i.e. of its first derivative, in fact. The remarkable passage which follows is nothing other than a transliteration of the graph represented in the diagram above: 'If the various properties of homogeneous bodies—e.g. the sweetness, yellowness and viscosity of honey—reach their full values according to the same law of mixture of the primary qualities, why does the change of one property not entail also the change of the others? For the colour of honey may change from yellow to white while its sweetness is not affected at all. On the other hand it is evident that the colour could not change without some change in the primary mixture of the honey. So why does taste not change along with colour if both obey the same law of the mixture of primary qualities? And similarly with wine: it can turn sour and thus change its taste without changing its colour, whereas both, if they obey the same law, should change with the mixture. And there are innumerable other examples. . . . Our explanation is as follows: each property is defined in a certain range, and is not given only at a single point. Whiteness is defined in a range (for when extreme whiteness is lost a body does not altogether cease to be white) and sweetness is also defined in a range (there is a wide range of sweet substances), and so on. If the properties have ranges, it is obvious that the efficacies of the mixtures from which these properties result must also have ranges. There must exist a certain extreme value for a mixture below which no property can come into existence, and for that given value the whole nature of the property will change at once. For instance, in order to exercise our minds on a paradigm, let ten parts of hot, cold, dry and moist be the amount which gives the full value of sweetness. If this amount is diminished by one for each of the primary qualities, sweetness will slightly decrease but it will not vanish. But if the amount is reduced by five (assuming that up to this point sweetness can be preserved), the whole property of sweetness will disappear. If, now, the different properties of honey result from the same mixture, they will not originate according to the same law of change; but if for instance sweetness has reached its maximum value, colour and viscosity may not yet have reached their maximum but may be much below it. Therefore, if the mixture is slightly varied, sweetness will not alter appreciably but colour may change completely, because it is near that particular and critical point of the mixture of primary qualities at which the colour in question can no more come into existence' [87].

Philoponus here gives a rather clumsy but unmistakable description of the dependence of two variables—sweetness and colour—on the independent variable 'primary quality'. This is the only case on record in antiquity of the quantitative treatment of a functional relationship, and it goes to show how the development of a concept of supreme importance for science was held up for lack of a suitable method of description—in this case graphical representation. Instances of functional thought of a qualitative nature have been known since the Stoics introduced their dynamic notion of a continuum.[15] There are examples of the description of a functional dependence of changes in the affected organs of sick people and changes of symptoms. There was further the description of the dependence of the seasons on the varying positions of the sun and, generally, the conception that a physical state, *hexis* in Stoic terminology, can run through a whole continuum of changes. There was even the recognition of the fact that a variable can possess an extremal value, exemplified by the straightness of a rod which represents a singular case within the range of all possible curvatures. But these beginnings were not followed up in later years, and Philoponus' example remained an isolated fact for the reasons already stated.

4. *Neo-Platonic revival of the geometrical theory*

The qualitative theory of matter, its preponderance acquired by the authority of Aristotle, the support of Stoic physics, and the backing of medical writers, could not obliterate completely the memories of the rival theories. Sometimes comparison with these was not very flattering to Aristotle's approach. Less than a hundred years before Galen, Plutarch in his essay on *The Principle of Cold* remarked that even if Plato's theory errs in particulars, its principles are built on a sound scientific basis because they go back to the substance of all things. He continues: 'This, it would seem, is the great difference between a philosopher and a physician or a farmer or a flute player; for the latter are content to examine the causes most remote from the first cause, since as soon as the most immediate cause of an effect is grasped . . . that is enough to enable a technician to do his proper job. But when the natural philosopher sets out to find the truth as a matter of speculative knowledge, the discovery of immediate causes is not the end but the beginning of his journey to the first and highest causes. This is the reason why Plato and Democritus, when they inquired into the causes of heat and heaviness, were right in not stopping their investigation with earth and fire but rather, they traced sensible phenomena to rational origins until they reached, as it were, the minimum number of seeds' [114].

During the whole of the Hellenistic period there must have been people who, like Plutarch, felt that the mechanistic as well as the mathematical approach to the problem of matter was leading to a deeper layer of physical reality than did the qualitative explanation. Going back to 'rational origins' was a desire that reflected the true classical Greek spirit of the Milesian philosophers, of the Pythagoreans and the atomists. It was a challenge which after a break of many centuries was met again by the Neo-Platonists. The revival of the mathematical theory of matter took place at a rather late period of Neo-Platonism and is connected with the names of Proclus and Simplicius; it was the last stage of a development that began with Plotinus and was continued by Iamblichus; thus altogether this system flourished for a period of three hundred years. *Prima facie* one could expect that a revival of Plato would also lead to a renewed interest in his ideas on matter. It is also known that Plotinus had a good knowledge of science, and we are informed on this point not only by his pupil and biographer Porphyry, but also directly in many passages in Plotinus' *Enneads* and by Simplicius who mentions that Plotinus joined Ptolemy in certain views on matters of physical science.[16] But the return to Plato came in a much more roundabout way and on a much more advanced level of insight. Indeed, the most fascinating aspect of the story is the very fact that Neo-Platonism which eventually became a spiritual and mystical system concerned with the salvation of man should have had such a weighty influence on the ideas of the nature of matter and of scientific explanation.

Plotinus conceived of matter in the widest sense of the word as the abstract receptacle of all corporeal events. Seen from this aspect, matter for him was essentially evil, an obstacle for the spirit struggling to achieve the good. Projected on the physical plane this intrinsic nature of matter reveals itself as indetermination, as something which cannot be fully grasped by our senses. Matter is 'the very non-being, an image and phantasm of mass, an aspiration towards physical reality, it is static but not in the sense of having position, it is in itself invisible and escapes those who want to observe it' [109]. Adhering to the old principle that like can be recognized only by like, Plotinus came to the conclusion that there must be an element of indetermination in the human soul if it is able to perceive something which by its own nature is indeterminate: 'How can I conceive the sizelessness of matter? . . . Only by indetermination. For if like is perceived by like, the indeterminate perceives the indeterminate' [107]. When confronted by matter, the soul experiences what Plotinus calls 'the impact of the shapeless'.

Here we are aware of the first germ of the Neo-Platonic antagonism to Aristotle's conception of matter and the way it can be
explained. At the other end of the long road, three hundred years
later, Simplicius expressed it in a very pointed formula, when
commenting on Aristotle's 'perceptible things must be explained
by perceptible principles': 'The principles of perceptible things
need not all be perceptible; for Matter, being a principle of perceptible things, escapes perception' [214]. This splendid epigram-
matic statement which has been corroborated to such a striking
extent by quantum theory, the modern theory of matter, has
an interesting counterpart in a religious conception: God, the
supreme formgiver and core of all *gestalt*, is himself formless and
escapes *gestalt*.

The Aristotelian elements in Plotinus' philosophy offered a
way out of the impasse to which the soul is brought by the elusive-
ness of matter. The statement of the evil nature of matter needs
qualification in so far as matter does not exist as a pure substratum
but always in conjunction with form, and this immediately determines the attitude of the soul towards it. In Plotinus' words:
'And just as matter itself does not remain shapeless but is always
shaped in the objects, the soul also being pained at the indeterminate, immediately casts over it the form of objects, almost
shrinking from being outside reality' [108]. Thus there exists a
chance for the perceiving soul to introduce some order and *logos*
into the description of matter: 'For neither masses nor extensions are the primary categories . . . but number and reason'
[111]. From a new vantage point of cognition, Plotinus reached
the old Platonic conclusion that mathematics can serve as an
instrument with which the indetermination of matter can be
overcome. We must remember that Plato had assigned to mathematics an intermediate and autonomous position as a separate
entity lying between the world of Ideas which is inaccessible to
us, and the particulars of the material world. The objects of
mathematics—as Aristotle explains in the *Metaphysica*—are postulated by Plato as a kind of mediator between the eternal substances of Form and the substance of perceptible bodies. The
essential characteristic of this mediator, that of being both accessible to our mind and eternal and permanent in essence, had given
Plato the faith in mathematics as a language suitable for the
description and explanation of physical reality. But it was only

Iamblichus, a few decades after Plotinus, who arrived at a real insight into the role that mathematics can play in the physical sciences and who not only recognized clearly that mathematics can go a long way towards describing the laws of nature, but also succeeded in expressing this idea in a lucid and brilliantly formulated language.

Iamblichus perhaps personifies more than any other personality of that period the strange mental climate of the third century A.D. and the three following centuries. It was a period in which obscurantist tendencies of all shades steadily increased their hold over the minds of religious sects, philosophical systems and practising scientists alike. Occult sciences flourished and magic and alchemy were in vogue. On the other hand, the same thinkers began to look at reality with a vision singularly sharpened by a perspective of centuries through which they were able to see the great classical achievements of the Presocratics and of Plato and Aristotle. Despite the impact of new religions and the overwhelming influence of Oriental mysticism, these thinkers consciously belonged to the ancient Greek civilization. The four centuries and more during which Stoic philosophy prevailed had helped to maintain that feeling of continuity and tradition. It is true that the written language, though still Greek, had changed considerably and had become cumbersome and repetitive in style, but it had also acquired a greater flexibility of expression and accumulated a richer vocabulary of terminology. Situations, old achievements and new possibilities were seen and formulated by a mature mind, and this maturity was oddly intertwined with obscurity and irrationalism.

All this, as said before, applies very conspicuously to Iamblichus who appears to us almost as a split personality, and some of his books, for instance the *Mysteries* and *On the Common Mathematical Science*, give a fair idea of the polarization of thought in his time. The latter book is of the greatest interest to the history of scientific ideas because here for the first time a clear conception is developed of the possible applications of mathematics to science, and the procedure of mathematical physics is described with great lucidity. In one passage Iamblichus projects his ideas backwards to the Pythagoreans: 'They found out what is possible and impossible in the structure of the universe from what is possible and impossible in mathematics, and they derived

47

the celestial revolutions from causal rules according to commensurable numbers, by defining the measures of heaven by certain mathematical laws and, generally, by establishing the prognostic science of nature through mathematics and by making mathematics a principle for all that can be observed in the cosmos' [60]. Obviously the term 'prognostic science' was coined to describe the methods worked out since the days of Hipparchus and Ptolemy that had proved so successful in predicting the positions of the planets. But Iamblichus goes far beyond this when he envisages mathematics as a signpost and guide on the way of the exploration of 'the structure of the universe'.

In another passage Iamblichus emphasizes the necessity for an individual treatment of every subject in science according to its nature: 'One cannot look everywhere for the same causes and equally one cannot expect the same precision in everything. But just as we divide the technical arts according to the underlying materials and do not search for equal precision in gold or tin or bronze nor in different kinds of wood, we must have the same approach in the theoretical sciences. For the subject-matter will immediately make a difference. . . . It is not likely that all these will have the same or similar causes, but in so far as the principles differ, the proofs must differ too' [61].

In one of the last chapters, Iamblichus gives an elaborate definition of the methods of theoretical physics which is truly astonishing for its intuition and power of expression. In spite of its length it is worth quoting this remarkable passage which begins with a rebuttal of Aristotle's principle of explanation: 'Sometimes, it is also the practice of mathematical science to attack perceptible things with mathematical methods, such as the problem of the four elements, with geometry or arithmetic or with the methods of harmony, and similarly other problems. And because mathematics is prior to nature, it constructs its laws as derived from prior causes. This it does in several ways: either by *abstraction*, which means stripping the form involved in matter from the consideration of matter; or by *unification*, which means by introducing mathematical concepts into the physical objects and joining them together; or by *completion*, which means by adding the missing part to the corporeal forms which are not complete and thus making them complete; or by *representation*, which means looking at the equal and symmetrical things among the changing objects from the point of view where they can be best compared with mathematical forms; or by *participation*, which means considering how concepts in other things participate in a certain way in the pure concepts; or by *giving significance*, which means by becoming aware of a faint trace of a mathematical form taking shape in the realm of perceptible objects; or by *division*, which means considering the one and indivisible mathematical form as divided and plurified among individual things; or by *comparison*, which means looking at the pure forms of mathematics and those of perceptible objects and comparing them; or by *causal approach* from prior things, which means positing mathematical things as causes and examining together how the objects of the perceptible world arise from them. In this manner, I believe, we can attack mathematically everything in nature and in the world of change ' [62].

There is no reason whatsoever, in the light of what I said before about Iamblichus, to doubt his authorship of this splendid exposition and to assume that he copied it from somebody else. These words could very well have been written by the same man who occupied himself with theurgy and practised alchemy. But one must of course remember that they would never have been written by Iamblichus or any other thinker of that particular period had not Plato, seven hundred years earlier, made the first statement about the place of mathematics in the process of cognition. Iamblichus himself alludes several times to the Platonic origins of his point of view, as in the following sentences where he again dissociates himself from Aristotle: ' . . . So are the mathematical sciences which I think have to be imagined as something like the Ideas and in them they have their foundation. They are not to be conceived as abstractions of the perceptible things, but being below the ideas they derive from them their image-like character by acquiring magnitude and appearing as extension' [59]. Iamblichus elaborates this theme by paraphrasing Plato's celebrated simile of the shadows in the *Republic*.[17] In the perceptible world, the four elements of matter throw their shadows on the perceptible, material objects. Just as these shadows are caught and come to rest on the surface of the sensible world, so in the intelligible world mass and extension come to rest on the plane of the pure mathematical forms. But for

49

mathematics, the principles of the physical world would never be caught as intelligible and well-ordered patterns.

It does not detract at all from Iamblichus' originality of thought that his conceptions are rooted in Plato, nor can his achievement be belittled by the fact that it remained a purely programmatic one. In the history of ideas the programmatic level is an essential step of a developing conception on its way towards practical realization. Those who succeed for the first time in arriving at the clear formulation of a programme may for some reason inherent in history or by sheer accident precede the fulfilment of that programme by centuries. Iamblichus' extraordinary scheme for theoretical physics was suggested in A.D. 300, and it had to wait for nearly 1400 years to see the beginning of its realization. Francis Bacon, who was not a creative scientist either, succeeded in formulating a programme for the experimental sciences whose realization was begun in the century in which he died. Unlike Iamblichus, he had the luck to live in a civilization which was in its prime, when tradition and continuity of thought began to establish themselves.

The story of the mathematical theory of matter continues with Proclus (c. 410–485), one hundred and fifty years after Iamblichus. In his commentary on Euclid he re-states and amplifies Iamblichus' exposition on the place of mathematics between the real and the corporeal being. Some of these ideas appear again in his commentary on the *Timaeus* which is only partly extant. Also among the parts lost are those relating to Plato's theory of matter. Fortunately, however, fragments of another of Proclus' books, entitled *Inquiry into Aristotle's objections against the Timaeus*, are preserved in extensive quotations in Simplicius' commentary on Aristotle's *De caelo*. Simplicius complements the verbal quotations by a fairly detailed report of the contents and by marginal comments and remarks of his own. Thus we are in the position of being able to follow an absorbing argument across the gulf of eight hundred years, an argument in which the last of the great Neo-Platonists takes issue with Aristotle over his criticism of Plato's geometrical theory of matter in the third book of the *De caelo*. Proclus, well-versed both in Plato and Aristotle, undertakes to refute Aristotle's objections one by one in a most methodical and careful manner. The many points he deals with are partly of a technical nature and partly touch upon matters of principle. Several of these passages will be quoted here because they are documents of the greatest historical interest in many respects. They show to what extent the tradition of the Academy had kept alive the spirit of the quantitative approach to the problem of matter. They further indicate that on certain, not unimportant points of interpretation there was a difference of opinion among the later exegetes of Plato. One gets a vivid impression of Proclus' attempts to make the most of the master's scanty and rather sketchy outlines and to put them on a firmer and more consistent foundation.

Simplicius, before beginning his account of Proclus' arguments, leaves us in no doubt about the poor opinion the Neo-Platonists held of Aristotle's qualitative theory: 'Plato derived the origin of the four elements, fire, air, water and earth, from principles which are more fundamental than the qualities hot, cold, dry and moist, namely from quantitative differences that are more adequate for the explanation of matter. This is evident from the fact that he accounts for the differences of those qualities by the differences of the geometrical shapes. We are told by Theophrastus that already Democritus has said before Plato that the qualitative explanation is a primitive one, because our soul feels the need to conceive of a principle more fitting for matter than that of the activity of heat' [213].

Earlier in this chapter a short summary was given of Plato's theory. One of Aristotle's first objections to it was that it does not allow for the change of earth into one of the other elements because of the difference in shape between the triangular constituent element of the cube, which represents earth, and that of the three other bodies. Proclus' answer to this is twofold. It first relates to the empirical conception of earth already to be found in Aristotle's *Meteorologica*, as a predominantly solid substance which, however, has a certain admixture of air or water, as for instance metals: 'Against this objection Proclus replies that one can reverse the argument and say that those who allow for a change of immutable earth do not adhere to the phenomena. Nowhere does experience show that earth undergoes a change into other elements; in fact, earthlike substances only change in so far as they are contaminated with air and water, but pure earth, for example ashes or dust, is completely unchangeable' [215].

The second part of Proclus' reply is a theoretical one and goes beyond what is written in the *Timaeus*. It is of the greatest interest because of its relativization of the concept of an element. Here even the so-called last units are compound entities, relative to the theoretical possibility of a still more advanced degree of decomposition. This conception introduces a kind of ladder of 'ultimate' units, and transformations which are not possible on one rung of this ladder may become so on a lower one: 'In so far as earth is made of primary matter Plato regards it as changeable into other elements, and only in so far as it is connected with the isosceles triangle it is unchangeable. Indeed, as long as the triangles conserve their specific character, earth cannot originate from the halves of the equilateral triangles nor can the other elements originate from the isosceles. But when the triangles themselves are broken up and reconstitute themselves again into shapes, a former isosceles triangle or a part of it can become half of an equilateral one. When the dissolution of the triangles is carried down to primary matter the mutual transformation of earth and the other elements is a plain fact. Otherwise completely unformed matter, able to receive the shapes of all things, would be an empty concept' [216].

Objections to Plato's 'chemistry' of the transformation of water into air and vice versa were made not only by Aristotle but also by Alexander of Aphrodisias. The latter asked a very apt question pertaining to the possible direction of a transformation. According to Plato water is transformed into air and fire according to the formula:

1 unit of Water (20 triangles) $=$ 2 units of Air (2 $\times$ 8 triangles)
$+$ 1 unit of Fire (4 triangles).

But what about the reverse process? Air (i.e. vapour) becomes water by cooling and condensation, and thus it does not make sense to assume that this process should give rise to the formation of fire, as is suggested by the formula:

3 units of Air (3 $\times$ 8 triangles) $=$ 1 unit of Water (20 triangles)
$+$ 1 unit of Fire (4 triangles).

Simplicius relates Proclus' reply as follows: 'The philosopher Proclus says that in the process of dissolving water into air, whereby fire is the resolving agent, two parts of air and one part of fire are produced. However, in the reverse process of air becoming water, three parts of air dissolve (forming one part of water) and the remaining four triangles, by the same process of condensation, combine with two other parts of air and together form one part of water' [217]. Expressed in a formula, this means the transformation of five parts of air into two parts of water in the following two stages:

Stage one: 3 units of Air (3 × 8 triangles) = 1 unit of
Water (20 triangles) + 4 free triangles.

Stage two: 4 free triangles + 2 units of Air (2 × 8 triangles)
= 1 unit of Water (20 triangles).

The assumption of an intermediate state where four free triangles exist for some time 'in suspension' was already criticized by Aristotle. But Proclus remarks: 'There is nothing strange in the fact that some unformed matter should be involved. In all transformations one has to allow for the existence of some shapeless matter for some time, whereby the resolved matter enters in some form into the resolving mixture. And one can hardly deny that in the process of metabolism in our body, too, certain parts of matter often remain in an intermediate state' [217].

The temporary 'suspension' of triangles in an unformed state during a process of transformation is a perfectly clear picture if considered as an interpretation of the formulae. But how far can one go in the literal conception of such pictures? The history of physics of the last three hundred years has shown that models fulfil a useful and even necessary function in the first stages of physical theories, but they have to be discarded gradually with the further development of a theory and the advancement of its conceptual content. The second half of the nineteenth century witnessed Maxwell's abandonment of all mechanical models of the electromagnetic field and later the beginning of the end of the aether hypothesis, after Michelson's experiments. Models were again constructed in the first stages of the quantum theory of the atom, and again discarded, and they reappeared later as auxiliary pictures for the understanding of nuclear structure.

To us, who have become used to the inevitable process of an ever-increasing abstractedness of scientific cognition, the discussions of some of the features of Plato's theory sound familiar. Plato, to judge by what he says in the *Timaeus* (53 c), did not

E

attribute corporeality to the triangles. Otherwise he would not have contrasted the 'depth' of a body, i.e. its tri-dimensionality, with the surface enclosing it. For Plato the triangles seem to have been the mathematical elements of the incorporeal surfaces governing the corporeal processes of matter. Aristotle took strong exception to this saying that a theory does not make sense if it derives corporeal generation from surfaces.[18] The opinions of the Neo-Platonists on this point were divided: Simplicius tells us that 'the physical explanation by geometrical figures is believed by some of Plato's interpreters, among them Iamblichus, to have been meant only symbolically; but the later Platonic philosophers attempt to show that it has to be taken literally' [210].

Proclus, curiously enough, belonged to the latter group, adopting a rather mechanistic attitude, but reasoning along Aristotelian lines. Simplicius informs us very briefly on this: 'Proclus replies to this, that the physical planes are not without depth. For if a body confers spatiality upon whiteness when the whiteness falls upon it, this will apply all the more to the planes encompassing the body. But if the plane has depth, bodies are not generated from the incorporeal but rather a more complex body from a simpler one' [218]. Simplicius seems to have inclined to the other view. He takes issue with Alexander who had asked in what respect Plato differed from Democritus, as both theories postulate shape as the characterizing element of physical bodies. Simplicius replies that, indeed, in this respect there is no difference between the two theories: 'However, Plato's theory differs in so far as it makes the assumption of the plane as something more simple and basic than the atoms, which are bodies, and that it sees in the shapes creative symmetries and proportions . . .' [212].

Aristotle had tried to show the incompatibility of Plato's atomistic hypotheses with his denial of a vacuum. How can Plato explain away the empty places and interstices between his cubes, pyramids and the other elementary bodies? Because of the *horror vacui* which prevailed so conspicuously among the Greek scientists—the Democritean School and Strato excepted—the discussion of this point takes up considerable space. Proclus reminds us that Plato had assumed different sets of elementary bodies, varying in size, and thus the smaller bodies can fill the gaps between the larger ones. Further, there is always some fire in air (and thus small pyramids closing the gaps between the octahedrons), and some air in water (thus small octahedrons filling the interstices between the isocahedrons). There is also no difficulty in explaining how water, for instance, closes tightly against the walls of a jar if one takes into account the possible interpenetration of the elements of the fluid and the container: 'For the surrounding body also consists of rectilinear shapes, and nothing prevents them from adapting themselves to each other. Our difficulty arises from the fact that we can see the cylindrical or spherical surfaces of the vessels and forget that they, too, are composed of rectilinear elements. The substance in the vessels and the vessels themselves consist of elements, and they all mutually adapt themselves according to the shape of the planes' [220].

It is interesting to observe how Proclus, taking his stand on the basic presuppositions of every atomic hypothesis, is having recourse to some of the old Democritean arguments. To us, composite matter appears homogeneous because of the smallness of the elements, 'as happens with colours when they are side by side in small patches; their mixture has the appearance of uniformity: for instance cloth whose warp and weft are of different colours' [221]. Proclus gives another technical illustration: glue, cementing together different materials, does not eliminate the individuality of the particles of which these materials are composed. And further: when many torches are carried together, their light merges into one single flame, but when the torchbearers disperse each flame and its light appear separately again. Similarly with the sound of a choir, whose different tunes merge into one harmony.[19] All these similes remind us of Anaxagoras and Democritus.

Of greater importance are the arguments of a more theoretical nature. Aristotle, completely ignoring the principles of atomism, had asked how a pyramid could be an element, since on division into two parts one part would not be a pyramid, and thus a pyramid could not be the ultimate unit. 'Proclus, replying to this, rebukes Aristotle for declaring the pyramid to be fire and for not remaining within the frame of Plato's assumptions who had said that the pyramid is a seed of fire rather than fire itself. For fire is an agglomeration of pyramids individually invisible because of their smallness. As long as fire is divided into smaller parts it will be divided into pyramids; the single pyramid however is no

longer fire but is an element of fire, invisible because of its smallness. A part of a divided pyramid is not an element nor does it
consist of elements, because a single pyramid cannot be broken
down into pyramids but into planes' [219].

This is straightforward thinking in atomic terms, and so is
Proclus' reply to another objection of Aristotle, namely that, if
what is burned is turned into fire, the object burned must turn
into pyramids, which is as absurd as maintaining that a knife cuts
things into knives: 'To this Proclus replies that fire certainly dissolves the elements of the burning matter and transforms them
into its own; but a knife does not act upon the primary matter of
the material cut; it does not dissolve it but merely divides it
quantitatively by turning a larger bulk into a smaller. Nor is the
shape of the knife determined by nature but is something accidental to it. If this is so, how could the knife produce other
knives by division?' [223].

Lastly, there are two arguments which more than any of the
others bring out the essence of the mathematical conception of
matter. Two of the Aristotelian objections go to the heart of the
explanation by geometrical symbols. He says with regard to the
pyramid representing fire that if the power of fire to heat and
burn lies in the angles of the pyramid, all the elements, in varying degrees, must have this power, because the bodies representing them all have angles. Further, how can shapes which have no
opposite represent opposite qualities like hot and cold? The first
objection is answered by Proclus as follows: 'The conception that
the angle itself is heating is wrong and leads to false conclusions.
The *Timaeus*, starting from perception, states that the condition
of hot is something acute and dividing. But something that cuts
does not do so simply because it has angles but because of the
acuteness of the angles and the thinness of the edges. On this
principle the crafts produce cutting tools and nature has sharpened the angles and thinned the edges of the cutting teeth, and
has broadened the chewing teeth. Speed of motion is also needed.
One can thus not attribute the faculty of heating to the angle
alone but to the piercing acuteness of the angle and to the slicing
thinness of the edge, and to the speed of motion' [222].

The reply to the second objection clarifies still more the point
in question: 'Proclus refutes this objection by saying that the
question of attributing a shape to the cold may be well put, but

that one has also to remember that the pyramid itself is not heat; heat is the cutting faculty which results from the acuteness of the angles and the thinness of the sides. Now cold in itself is not a shape, as neither is heat, but it is a faculty of a certain shape. And like heat which is given by the acuteness of the angles and the thinness of the sides, cold on the contrary is given by the bluntness of the angles and the thickness of the sides. One faculty is thus opposed to the other; it is not the shapes which are opposites but the faculties which reside in the shapes. The theory thus requires not an opposite shape but an opposite faculty' [224].

These two replies offer, in a nutshell, the final stand of the Platonists against Aristotle, of the mathematical theory of matter against the qualitative: in the last instance, at the ultimate stage of explanation, one has to return to the qualities, but the way to this leads through mathematics. The qualities of which Proclus speaks are not the perceptible qualities of Aristotle, they are the qualities of geometrical shapes symbolizing the elements of matter. This is in fact the principle of explanation of physical phenomena which forms the basis of theoretical physics today and which has gradually evolved since the days of Newton. The theoretical physicist operates with physical quantities, concepts constructed from the elements of physical reality and having their roots in empirical phenomena. These physical quantities are identified with mathematical symbols or quantities and enter into mathematical equations which express certain laws or regularities of nature. The mathematical relations and quantities forming the final result of these equations, corresponding in their turn to certain physical quantities, are then translated back into the qualitative language of perception.

The programmatic anticipation of this approach to the problem of nature can be clearly recognized in the Neo-Platonic writings, and very conspicuously so in Proclus' elaboration of Plato's geometrical theory of matter. It is true that it was a crude and rudimentary anticipation, lacking the proper mathematical tools and practically without any sound experimental foundation. But seen historically against the background of the two rival theories, the purely qualitative and the mechanistic, the Neo-Platonic revival of the mathematical approach stands out as one of the great contributions of late antiquity to the history of scientific ideas.

One should not overlook a psychological factor which no doubt

helped to stimulate the interest of the Neo-Platonists in a mathematical theory of matter. The progress of mathematical astronomy, the great achievements of Hipparchus and Ptolemy, had
proved that man can succeed in 'saving the phenomena' by
mathematical analysis, and that mathematics, in an ever-
increasing measure of precision, was an instrument for the pre-
diction of astronomical events, of future positions of the sun, the
moon and the planets. Mathematics had transformed astronomy
into what Iamblichus called a 'prognostic science'. This created
the psychological situation of deep dissatisfaction with the explanation of matter in which Aristotelian concepts still prevailed.
The failure of the qualitative approach was made still more conspicuous by the success of quantitative methods in astronomy.
Why could not some plausible mathematical assumptions be
made with regard to matter, as had been done in astronomy?
There are passages in Simplicius which prove beyond doubt that
these were the origins of the Neo-Platonist desire for a quantitative attack on the problem of matter. The following quotation may serve as an illustration:

'All this I have put down here in order to show that it was not
without reason that the Pythagoreans and Democritus proceeded
to geometrical shapes in their inquiries into the principles of
qualities. But it is possible that the Pythagoreans and Plato did
not postulate the triangular constitution of things as something
absolute. Their procedure may have been like that of the various
astronomers who made hypotheses based on the firm belief that
the diversities in heaven are not what they seem to be but that
one can save the phenomena by making the basic assumption of
a regular and circular motion of the celestial bodies. Similarly
the Pythagoreans, preferring in principle the quantitative to the
qualitative and shape to quality, choose as elements of the bodies
those geometrical forms which are more in the nature of a principle and which are superior for reasons of similarity and symmetry, and which they regarded as sufficient to account for the
causes of physical events' [211].[20] Here, for the first and last
time in antiquity, we find astronomy and the theory of matter
mentioned together in the same context as two fields of science
whose task it is to use mathematics as a language for the explanation of physical phenomena.

The close affinity of the Neo-Platonic attitude to that of our present age can be illustrated by a comparison of Simplicius' words with the following quotation: 'Our experience hitherto justifies us in believing that nature is the realization of the simplest conceivable mathematical ideas. I am convinced that we can discover, by means of purely mathematical constructions, the concepts and the laws connecting them with each other, which furnish the key to the understanding of natural phenomena.' This was said 1400 years after Simplicius by Albert Einstein.

Finally, a few words should be said about the role of alchemy in these developments. Despite the fact that Greek alchemy began to grow during the Neo-Platonic period and was practised by many Neo-Platonists, there is no conclusive evidence as to any interaction of importance between alchemical conceptions and the lines of theoretical thought described in this chapter. Greek alchemy developed a practice, a system of associations and a language that bound it up strongly with the world of magic and astrology and made it impervious to rational considerations and to mathematical methods. We do not know whether Proclus, like Iamblichus before him, was a practising alchemist; a few passages in his writings show that he was at least theoretically interested in alchemy, but he did not allow his alchemical interests to interfere with his inquiries into the structure of matter, in any case not to an extent that could have induced Simplicius to mention it in his extensive account of Proclus' book against Aristotle.

The picture of Proclus, the scientist-philosopher, would not be complete without a short reference to his alchemical passages. In his commentary on the *Timaeus*, he refers to what Plato says at the beginning of his book on the training of the guardians of the city, that they should not think of gold and silver, or any other possession as their own private property.[21] 'If one wants,' remarks Proclus, 'this can be explained from the point of view of physical science. Gold and silver and every metal, like other substances, grow in the earth under the influence of the celestial gods and their emanations. Gold is attributed to the Sun, silver to the Moon, lead to Saturn and iron to Mars. These metals have their origin in heaven but they exist in the earth and not in those which emit these emanations. For nothing involved in matter is admitted in heaven. And though all substances originate from all the gods, there is yet in everything another specific prevalence, some belonging to Saturn, other to the Sun; the men who are

59

fond of contemplating these matters compare these and attribute to them various faculties. These substances are thus not the private property of gods but they are common property—for they originate in all of them, nor do they reside in them—for the active powers do not need them, but they are compounded on earth through the influence emanating from the gods' [116].

All this is said in a truly alchemistic spirit, and so are Proclus' remarks when speculating on the significance of Plato's words about the invisible rivets by which the particles of matter are welded together.[22] Besides using certain technical terms which are to be found also in such well-known alchemistic writings as the Stockholm and Leyden papyri, he says, apparently quoting an alchemistic epigram: 'Everything is dissolved by fire and is glued together by water' [122]. There is however one place where Proclus mentions astronomy (not astrology) in the same context as alchemy, and that is in his commentary on the *Republic*. It is a passage characteristic of the Neo-Platonic belief in the essential wholeness of nature, a wholeness which can never be fully understood by procedures involving fragmentation and decomposition into artificial parts. Proclus begins by remarking that the astronomers try by mechanical means to explain the irregular motions in the heavens as composed of regular and circular ones, and the calendar-makers attempt to imitate nature which had created everything before they began their calculations: 'And there are those pretending to make gold out of the mixture of certain species, while nature makes the one species of gold before the mixture of those species of which they talk. And everywhere we see the same attitude, that the human soul hunts after nature with skilful devices to find out how things are generated. With regard to the stars there is a purpose in this which, not by chance, has given men success in their inquiries into the regular motions of the bodies moving in circles; for this, they say, is fitting to divine bodies' [123]. It rather looks as if Proclus had his doubts about the alchemists and their practices. His positive remarks about the harmonic analyses of the astronomers, in spite of his reserved attitude towards their method (of which more will be said in the last section of Chapter V), show his high regard for the practical achievements of astronomy.

One has to bear in mind the decidedly mystical taint of alchemy, the degree to which the results of its processes were believed to be dependent on astrological constellations, or even more so on the psychological state of the performer of the experiment and, generally, the strong anti-rational physiognomy of its practice. In so far as it was directed towards technical ends, Greek alchemy consisted mainly of a collection of hard-and-fast rules and recipes without any attempt at systematization or critical analysis. In all its essential features it was the opposite of scientific thought whose conspicuous characteristic, until the end of antiquity, was rational speculation at its best and an attempt at objectivity in the few cases where recourse was made to experience. When confronted by these two facts, one will not find it strange that there was no association of ideas to establish communication between the two compartments of thought in Proclus—that of the geometrical theory of matter and that of alchemy.

III

SUBLUNAR MECHANICS

---

1. *The laws of motion*

ARISTOTLE'S laws of motion occupied a central position not only in his own physics, but also in the subsequent discussion, interpretation and criticism which took place during the entire ancient period. Their importance in antiquity compares with that of classical mechanics in modern physics; Galileo's studies of acceleration and the laws of falling bodies inaugurated the era of modern science, while Newton's three laws and their application to planetary motion established classical mechanics. Further, Einstein's theory of relativity followed a renewed epistemological examination of these laws during the nineteenth century.

In a similar way, Aristotle's dynamics was in the first instance a serious attempt to fit a variety of motions into some rational framework and to regard them as simple expressions of order and harmony in nature. However, these laws turned out to be more than mere descriptions of some classes of motion. They reflected in their basic conception the major principles of a philosophy which governed all aspects of intellectual life until the beginning of the modern era. The fundamental distinction between sublunar and celestial events, the dichotomy between heaven and earth, revealed itself most succinctly in the contrast between the eternal and unchanging revolutions of the celestial spheres, and the motions in the terrestrial region which are finite and decaying whether 'natural' or 'forced', since in either case they have a beginning and an end. The 'natural' motion of a heavy or a light body, starting somewhere above the surface of the earth, leads the body in a straight line either up or down until it comes to rest in its natural place, unless prevented from doing so by other bodies. The two types of forced motion are exemplified by the throwing of a missile and by the pushing or dragging of a load. In the former case, the motion is originated by an impulsive force in a direction other than that of the natural motion of the missile and finally the missile comes to rest again in its natural place. In the latter situation the motion lasts only as long as the force is applied.

Aristotle's classification of bodies into heavy and light also conforms with an old Greek belief—expressed mainly by the Pythagoreans—in the interplay of 'opposing' principles or opposites in nature. Earth and Fire respectively are the heavy and light bodies *par excellence*; Water and Air, too, are heavy and light, but they occupy somewhat intermediary places between Earth and Fire. Moreover, one can change into the other by evaporation or condensation. Here we have a special case of another universal principle, embodied in the Aristotelian categories of 'potential' and 'actual'. Water is actually heavy, but it is potentially light and actually becomes so when it evaporates; and vice versa for Air. But potentiality and actuality, the main pillars of Aristotelian and scholastic philosophy, enter into the explanation of motion in a still more significant way. Light bodies move upward and heavy bodies downward in order to reach the state of their full actualization; this goal is attained as soon as they have arrived at their natural places. It is only there that earth *is* heavy and not *becoming* heavy, that air is light and not becoming light; and only there can they come to rest because they have reached the completion of their form—lightness or heaviness. Seen from this point of view, natural motion can be described as the realization of still another Aristotelian principle of supreme significance—the teleological principle. The natural motions of the elements, straight up or down, are tendencies towards their natural places, expressions of their striving towards the fulfilment of their proper form. The free-falling body is thus but another instance of the desire of unformed matter, of potentiality, to reach the actuality of its form, as in the case of the seed becoming a fruit-bearing tree.

These few remarks may suffice to indicate the pivotal significance of Aristotle's dynamics in understanding the interconnection of a systematic scientific theory and the basic philosophy of the age. This relationship is similar to that which developed in classical mechanics since the seventeenth century, its becoming interwoven with the epistemological approach of the age of Galileo and Newton and of the two centuries following. However, when it comes to quantitative statements, the characteristic differences in the scientific methods of the two ages—as is well known—were decisive in the whole course of development: in antiquity, predominantly crude approximations or incorrect formulations, and in modern times precise measurements and a careful and gradual establishment of a conceptual framework based on experience. Two examples are usually quoted as illustrations: Aristotle's assertion that the velocity of falling bodies is proportional to their weight contrasted with Galileo's laws, carefully established by the use of clock and yardstick; and the ancient belief in the proportionality of the moving force and the velocity of the moved body compared with Newton's laws of inertia and force. However, the antithesis stated in this way is too harsh, and the judgement needs qualification in at least two respects. The first is, that Aristotle himself was already aware of certain discrepancies between his quantitative formula and experience and tried to explain them rationally within the whole conceptual setup of his time; the second is that, as we shall see, considerable criticism was voiced in the post-Aristotelian period and late antiquity of many of Aristotle's statements. This shows that the whole difficult problem of dynamics was continually being approached from different angles because of the clear feeling of the need for a thorough revision.

One of the main reasons for the failure of antiquity to discover the correct laws of dynamics was that in establishing relations between forces as causes of motion and the resulting motions, no account was taken of the opposing forces of friction. It is strange that such obvious factors should have been overlooked at a time when the only motive force available was the muscular power of animals and men and when the devices used for reducing friction in the transport of large stones and other building materials must have been common knowledge. However, one should not forget the considerable conceptual difficulties in ascribing to friction the character of a force at a time when forces were associated with muscular strain and human exertion. Whatever the reason may have been, it led Aristotle, in his only quantitative formula, to assume proportionality not between force and acceleration but between force and velocity. This is the content of the relation expounded in the last chapter of book seven of the *Physica* (and repeated in another context in the third book of the *De caelo*). It states that the distance through which a load is moved by a force is in direct proportion to that force and to the time the force acts and in inverse proportion to the magnitude of the load. This indeed is equivalent to the (incorrect) statement that the force is equal to the product of load and velocity, in obvious contradiction to Newton's second law, in which acceleration takes the place of velocity. Aristotle's formula was in agreement with the belief common in antiquity that there is motion of a body only as long as a force acts on it and that the body comes to rest as soon as the force ceases to act.

In fairness to Aristotle, it is important to remember that he himself realized that his formula was an approximation valid only for large forces. As an illustration, he used the case of a gang of ship-haulers pulling a boat. If a certain number of men can haul a ship over a certain distance in a certain time, one cannot expect —as he states in somewhat cumbersome language—that one could reduce the hauling force *ad libitum* and still transport the ship over the same distance, or even a smaller one, in a proportionally longer time. In fact, he says, without indicating any reason, that the ship may not move at all. The reason which almost certainly was not known to Aristotle is, of course, that a minimum force is needed to overcome the friction of a body at rest, and that this friction is generally greater than that of the same body in motion.

I have repeated at some length details well known to historians of mechanics in order to throw into relief the significance of remarks and comments on this matter which were made in a later period and which show, at the same time, both scientific progress and some limitations inherent in the ancient approach to the facts of inorganic nature. Here is a passage by the commentator Themistius (about A.D. 320–390 in Byzantium) in which he comments on the problem just mentioned: 'There is a whole theory and problem involved in the question of whether several people together are able to move a load whose magnitude is the sum of what each of the individuals alone can move. I mean that if each can move a weight of a talent, then a hundred together could move either a weight of a hundred talents or more or less. That it should be less does not seem reasonable; more is rather probable, since the togetherness, ambition and mutual stimulus have their effect, as in the case of horses harnessed together for speed, whose heat and power increase through the intensity of their competition. On the other hand, it is quite reasonable, and even necessary, if the former assumption is true, that if the load is divided so that each gets his share of the whole, he will by no means be able to move it. For when the force is split up, something is lost which was added and increased by the togetherness. This, to a great extent, seems to be similar to the case of geometrical figures, where a double length yields a fourfold area, whereas in our case, all together are many times stronger. Loads which are moved also behave differently, depending on their shape, as do bodies that move by themselves. For what is spherical moves easily, and those bodies which are least stable and have the smallest surface of contact are easily furthered in their motion. Generally it is easier to further the motion of a moving body than to move a body at rest. This is likewise the case with all similar things' [232].

The last three sentences of this passage contain, in a sort of afterthought, the first general observation on record of the influence of friction on a body either in motion or at rest. It is of course of little importance whether these remarks are original or whether they refer to things already said by others. What matters is the context in which they are said here, as a scientific explanation of the relevant law stated by Aristotle, introducing friction into dynamics, if only in some qualitative way. The greater part of the passage, however, is of interest in quite another respect, in that it demonstrates conspicuously the anthropomorphic or zoomorphic approach of the ancient Greeks to physical problems which is the opposite of our tendency to transform biology into a chapter in physics or chemistry. Perhaps one could argue in this special case that explanation in terms such as competition or ambition is not so far-fetched because of the understandable identification of moving forces with animals or men. But there is ample evidence for the general belief that in inorganic nature, too, 'togetherness' increases the power of the whole more than in linear proportion to the number of its units. This belief was even carried to the absurd conclusion, contrary to experience, that weight must obey the same rule of non-additivity. One may say, though, that there is a certain inner logic in this belief, for in Greek physics weight was the innate force of a body producing its natural motion towards its natural place at the centre of the earth. In this respect the weight of a body was often compared to the soul of man. In the same way as man moves and acts by virtue of his soul, i.e. his *eidos* (form), a heavy body moves downward by virtue of its own weight, which is nothing other than its own *eidos*. This conception can be taken as the background of the following strange passage from John Philoponus: 'Qualities become more powerful when parts of the same form act together. This is obvious and can be seen more clearly in the case of weights. When you join together two weights of a pound, the combined weight will be heavier than the sum of the two, for it will not be two pounds, but more. Similarly, when you divide a weight of two pounds into two equal parts, each will not be a pound, but less. Thus things of the same form coming together become more powerful, and being divided they become weaker' [69].

Philoponus makes still another highly suggestive comment on the problem of the non-additivity of the forces of individuals when combined in a group. That comment refers to a short remark made by Aristotle in the same passage in which he states his law of dynamics. The gist of this remark, as interpreted by his commentators, is that individuals, on forming a group, lose their individuality in a certain sense. They forgo their actual individuality, and as parts of the group, have only potential existence. The actuality of the individuals is replaced by the actuality of the group as a whole. If a single part of the group ceases to exist, its share in the total effort cannot be estimated by assuming this effort to be the sum of contributions of discrete individuals. Philoponus adds a simile of his own: 'The part by itself, says Aristotle, does not actually exist within the whole. It does not act as a mover within the whole as it does on its own, when it has a well-defined individuality; it only has potential existence within the whole, and therefore must be regarded as unformed matter. It is a case similar to that of the parts of a noun of which Aristotle says in the book *On Interpretation* that they by themselves have no significance, but that each of them, being potentially in the whole, contributes to the full meaning of the noun. Similarly one ship-hauler by himself cannot move anything, but, being a potential individual within the whole, he adds to its motion with the rest' [82].

The influence of Stoic ideas on the relation of the whole to its parts is clearly shown in the example of the association of ship-haulers in a group or as individuals, and of the noun and its parts. The Stoics defined a whole hierarchy of structures beginning at the lower end with a simple agglomeration of individuals and rising to the structure of an organism where every part exists only in constant interaction with the whole. The whole, when considered as an organism, is more than a simple addition of its parts; this applies also to the noun and its different letters or syllables which receive semantic significance only in the context of the whole word. It applies equally to the ship-haulers whose common purpose and singlemindedness bind them into one organic whole which is more powerful than the simple sum of the forces of all the individual ship-haulers. All of this is typical of the Greek and Hellenistic way of thought; even more typical is the fact that it is stated as an interpretation of and a marginal comment on a chapter in physics—the basic laws of dynamics.

The very fact that as basic a law as that of Aristotle did not have universal validity but was restricted, as its author had already admitted, to a limited range of forces and loads, was rather disturbing for all students of this problem. Themistius' notion, that the problem was a more complex one and that other factors besides those introduced by Aristotle had to be taken into account, was not pursued by later commentators. However, they continued to worry about the obvious conclusion that there could be practical limitations of a physical law which theoretically could be expressed in precise mathematical terms without any apparent restriction whatsoever on its validity. That such a question was raised in connection with the first physico-mathematical equation ever established, is interesting in itself, notwithstanding the fact that for reasons well known to us that equation was not correct. Is an equation between physical parameters valid for every numerical value of each of these parameters? Of course, infinite values have to be excluded, for physics deals with finite quantities, as Aristotle had already explained at length in the third book of his *Physica*. But discrepancies between experiment and theory for finite values of a parameter, small or large, must inevitably disclose the approximate character of every physical law and lead to a modification or extension of theory. Boyle's law is an often-quoted example from modern times: as soon as large pressures could be applied to a gas, the limited validity of the assumed inverse proportionality of pressure and volume came to light, and the law had to be modified and the concepts of ideal and real gases were introduced. If experiments involving velocities comparable with that of light had been technically possible before the advent of the theory of relativity, deviations from the proportionality between force and acceleration would have raised the question of the limitations of Newton's second law and the physical reasons for them.

In view of this, it is worth quoting the comments of Simplicius, the contemporary of Philoponus, if only because of the emphasis given to the problem and its comparison with a celebrated law of statics where linear proportions were involved—Archimedes' law of the lever: 'It is worth while looking for the reason that proportionality between force and load is preserved, for example, in the case of the force being halved, but not for the whole range of force. . . . It is by no means clear why proportionality does not prevail for all values, but, should somebody assert that it does prevail, the still more paradoxical result would follow that one single man could move mount Athos if it happened to be separated from the earth. For if he could move one stone of Athos a certain distance in a certain time, why should he not be able to move the whole mountain through a much smaller distance in a much longer time? On the basis of this proportionality between force, load and distance, Archimedes devised the lifting machine called *charistion*, asserting that it would work proportionally for all values, and uttered his boastful "give me a place to stand on, and I will move the earth". It must now be said, as was briefly stated before, that not every force can move every load, no matter how small the distance or how long the time. . . . There exists a lower limit for the force, below which it cannot move any load through any distance whatsoever, and an upper limit for the load, above which it cannot be moved over any distance by any of the corporeally moving forces. In between there exists a range of conditions between forces, loads, distances and times such that

F

proportionality holds. . . . It is thus clear that there exists proportion as well as lack of proportion in the relation between forces that can give rise to motions and loads that are able to be moved as well as between distances and times' [164].

2. *The impetus*

The incorrect assumptions implied in Aristotle's law of dynamics, that proportionality holds between the moving forces and the velocity of the moved body, is equivalent to the assertion that motion can be maintained only as long as a force acts on the body. If the force vanishes, velocity will vanish too, that is the body will come to rest. Even for Aristotle himself, it was difficult to explain away the obvious incompatibility of that conclusion with experience for one type of 'forced' motion, namely the case of a missile thrown in a direction other than that of its 'natural' motion. Why does the missile continue to move upward after the force of the projecting sling, bow or human arm has stopped? What other invisible force continues to push the missile until it apparently fades out and the natural innate tendency of a heavy body forces the projectile downward on its natural path? Chapter ten of the eighth book of the *Physica* records the rather pathetic efforts of Aristotle to reply to these questions. He tries desperately to picture a mechanism of successive pushing forces maintaining the motion of the missile. In conformity with his basic conception that motion is only possible in a material medium (e.g. air or water), Aristotle assumes that contiguous parts of the medium, for example air, push the missile and maintain its motion. But the difficulty still remains: if mover and moved act and cease to act simultaneously, the picture of a spatio-temporal succession of pushing masses of air behind the moving missile is not tenable. 'We are forced, therefore, to suppose that the original mover conveys to air . . . the power of being a mover, but that the air does not cease simultaneously to be a mover and a moved; it ceases to be moved at the moment when its mover ceases to impart motion to it, but it continues to be a mover and so moves whatever is adjacent to it, and of this again the same thing is true' [18]. This is an extremely important passage, because Aristotle, in looking for a mechanism of transmission of motive power, proposes an hypothesis which contains the first germ of the celebrated idea of 'impetus' that had such a long subsequent history. He assumes that, while the intermediate agent ceases to be moved, or in other words, ceases to be acted upon by a force, it is still able to impart motion to its neighbour, thereby maintaining the chain of successive transmission until it gradually fades out. By making this assumption Aristotle did not adhere to the rigid conception that a force can act only by immediate contact and not by 'proxy', and initiated a stream of thought leading first to the concept of impetus and then to the modern notions of momentum and kinetic energy.

The first scientist to introduce the concept of impetus was the astronomer Hipparchus, two hundred years after Aristotle. Before discussing his theory, we will look at the comments made by some of Aristotle's commentators on the passage just quoted. Alexander of Aphrodisias (about A.D. 200), the most Aristotelian of all the commentators, remarks in a lengthy passage preserved for us by Simplicius: 'The air receives from the thrower of the missile the origin and key-note of being moved, as well as that of being able to move; the power it gets from the thrower is such that, being self-moved, it is able to be a mover. In some way it becomes for a short while a self-moved thing whose nature it is to absorb through some affection the power of the mover' [176]. Alexander's interpretation of the air that is moving without being moved as being self-moved is an implicit admission of the existence of some kind of 'impetus', of some faculty which is stored in the moving body. He later gives an illustration which is characteristic of the Peripatetics who conceived motion as a more general phenomenon, including change of size and of quality, as well as change of position. Fire, he says, is an active quality which can transfer some of its power to water. Water itself, after having been heated by fire, acquires the faculty of heating and keeps it for some time. This idea of storage of power is brought out still more clearly by Themistius, who followed Alexander; it is contrasted with the relationship between a magnet and a piece of iron in which the iron loses the temporary faculty of attraction as soon as it is separated from the magnet: 'Perhaps the neighbouring air is not only moved but also acquires for itself the power of moving, in direct analogy, I believe, to the case of a material that is heated by fire. It not only becomes hot but also acquires a power of its own to heat and passes this on continuously for some time. After a while this comes to an end when the power borrowed from the fire fades out in the process of transfer. Similarly, air and water . . . become, so to say, self-moved and thus, for some time, are both moved and moving simultaneously. However, they are not moved by the thrower, but rather by their own power which they received as a signal from the projector, exactly as water that was heated by fire not only remains warm after the fire has been removed, but conserves for a long time the power of heating' [233].

Still more articulate and distinct is Simplicius: 'Just as the motion of the thrower's body moves the air, so the air, while the motion remains in it, moves the air next to it and so on, until through the multiple transmissions the motion fades out after some time. There is other evidence for the continuance of motion for some time after the moving force has stopped, in cases where the moving power is considerable and the body moved is of suitable shape: the top which continues to rotate long after it has been set in motion by spinning, and the cymbal which, once having been struck, continues to sound for a long time because it remains in motion and keeps the air moving. These objects were set in motion by a mover and by their favourable disposition to move remain in motion even after the mover has been withdrawn. It is to this that Alexander referred when he said 'they become in some small measure self-moved', for the moved body remains in motion and induces motion in the body next to it' [177].

Simplicius finally sums up the argument by coining a technical term for the mechanism suggested by Aristotle: 'It is thus impossible to solve the difficulties of the problem of the thrown missile otherwise than by supposing that the moved object also receives kinetic power from the moving one' [179]. For a moment it seems as if Simplicius were ready to take the final step towards the concept of impetus (actually taken by his contemporary Philoponus and nearly 700 years before them by Hipparchus), when he puts the question: 'But if we say that the man throwing the missile transfers to the air a steady motion, why don't we say that this motion is given to the missile without having recourse to the air and therefore without our being forced to assume that it is not only moved but also moving?' [178]. However, Simplicius fails to draw the final conclusion and to eliminate the air completely from the picture. He tries to settle the question by orthodox

72

Aristotelian arguments and to explain the necessity for an intermediary action of the air by saying that as an intermediate element that can move in all directions, it has a certain stabilizing faculty. He thus remains, together with Aristotle, Alexander and Themistius, only a precursor of the concept of impetus although one clearly recognizes in all these thinkers a line of progressive conceptual and terminological clarification in the right direction.

The first scientist on record to have a conception of impetus, i.e. of a vectorial quantity which is imparted to a moving body by the instantaneous action of a force and then stored in the body, was the great astronomer Hipparchus (*c.* 190–120 B.C.). It must be admitted that the evidence is scanty, but the extant fragment referring to the throwing of a body vertically upward hardly allows of any other interpretation. The main features of the vertical throw were, of course, known much earlier—a slowing-down of the missile on its way upward followed by the accelerated motion of the freely falling body. Hipparchus, involved in a conceptual argument on the problem of weight and falling bodies which will be discussed later on, had an explanation of his own for the vertical throw, as recorded by Simplicius: 'Hipparchus in his book *On Bodies Carried Down by their Weight* says that, in the case of earth thrown upward, the throwing force is the cause of the upward motion, as long as it is stronger than the power of the thrown body; the stronger the throwing force, the swifter the object moves upward. Then, as the force diminishes, the upward motion continues with a reduced velocity, until the body starts to move downward under the influence of its own natural pull although the projecting force persists in a certain way; as this fades, the body moves downward more swiftly, achieving its greatest velocity when that force has completely disappeared' [203].

In these few lines we have a clear exposition of the concept of a 'throwing force', corresponding to the ideas of Philoponus in the sixth century A.D. and of some Arabic scientists in the twelfth century, and to the impetus theory of Ockham, Buridan and Oresme in the fourteenth century. There is no mention of continuous contact of a moving force with the moved body, or of any intermediary role played by the air as a mover. The force of the thrower confers upon the missile an impetus which gradually fades away. But here Hipparchus develops another original idea: the impetus has not completely died away when the body has reached its highest point of ascent; it continues to act during the descent, the acceleration in the downward motion being nothing other than the result of the gradual fading of the impetus, i.e. of a vector which is pointed upward and whose gradual diminution brings the natural downward pull into full effect.

The same explanation of the acceleration of gravity as a progressive weakening of the residual impetus was also given by some Moslem scientists, but was rejected by others who pointed out that the theory cannot hold in the case of a body whose fall is not preceded by a throw upward. However, Hipparchus extended his theory to this case, as shown by the continuation of the passage quoted above: 'He assigns the same cause also to bodies dropped from above. For in these also the force which holds them back continues for a while, and its opposing action is the cause of the slower movement of the falling body at the beginning' [203]. What Hipparchus must have had in mind is a kind of 'potential impetus' residing latently in every body that is not in its natural place (i.e. the centre of the earth in the case of a heavy body), regardless of the manner in which the body arrived in its present position; for no matter what its history before the start of its fall, the very fact that the body is at rest at a distance from its natural place is equivalent to its having an upward-pointing impetus imparted to it which is gradually dissipated during the falling motion. Unlike the modern notion of potential energy which is a scalar term, Hipparchus' potential impetus is a vectorial quantity directed away from the centre of the earth.

We do not know how Hipparchus arrived at the conception of impetus, nor what considerations led him to drop the Aristotelian theory that air sustains the 'forced' motion of a missile; if he actually refuted that theory, his arguments have not come down to us. The historian has to put on record that this idea was originated by Hipparchus—a fact of considerable importance in view of the eminence of Hipparchus as an astronomer and scientist. However, priority of scientific ideas has little significance in antiquity where continuity of thought and scientific tradition existed only sporadically. It was not until almost 700 years after Hipparchus, in the first half of the sixth century A.D., that the idea of impetus was developed again by John Philoponus whom we have to regard as its rediscoverer, considering the many centuries during

which the Aristotelian theory remained in vogue, and taking
into account Philoponus' elaborate refutation of that theory.[23]
Philoponus analyses two mechanisms which had been suggested
for the alleged function of air in sustaining the motion of the
missile. One was the commonly accepted notion of *antiperistasis*,
the interchange of places, already rejected by Aristotle. It was the
idea that the air in front of the moving missile is constantly being
pushed to the rear of it and there presses on the missile in the
direction of its motion and thus acts as a moving force. After
having disposed of this theory by a number of very plausible and
forceful arguments, Philoponus turns to the Aristotelian hypothesis that together with the push given to the missile by the
original thrower, the air behind the missile is set in rapid motion
and continues to push the missile: 'First we must ask those who
hold this view the following question: If somebody throws a stone
by force, is it by pushing the air behind the stone that he forces
it into motion against its natural direction? Or does the thrower
impart to the stone a kinetic power? If he does not impart any
power to the stone, but moves it merely by pushing the air, and
similarly the bowstring the arrow, what is the use of the contact
between the hand and the stone or the string and the notched
end of the arrow?' [73]. Philoponus now suggests an imaginary
experiment: Suppose a powerful machine generates a stream of
air much stronger than the throwing force—would the missile be
set in motion and be moved in the observed way? The answer is
in the negative. 'And further, if string and arrow or hand and
stone are in direct contact, and there is nothing between them,
what air behind the missile is being moved? And if the air at the
side is moved, what connection has it with the thrown body, for
it is not within its path? From these and many other considerations one can see that it is impossible that things moved by force
should move in this way. It must rather be that some incorporeal
kinetic power is imparted by the thrower to the object thrown
and that the pushed air contributes either nothing or very little
to this motion. If, then, objects are moved by force in this manner, it is evident that if an arrow or a stone is projected by force
and contrary to nature in a vacuum, the same thing will happen
much more easily, nothing being necessary except the thrower'
[74].

The essence of the impetus is brought out in this passage with the greatest clarity. It is a 'kinetic power' transferred at the moment of throwing from the thrower to the object thrown, by virtue of which it is kept moving in its 'forced motion'. The medium, e.g. air, does not help; on the contrary — the missile would move more easily in a vacuum. Philoponus calls the impetus 'incorporeal'. The Stoics would certainly have objected to this, because they regarded every quantity capable of physical action as 'corporeal'. Obviously Philoponus wants to emphasize the difference between his 'kinetic power', which is akin to our vectorial term 'momentum' or the scalar term 'kinetic energy', and a material agent like air, constantly pushing the missile. In the continuation of the passage quoted he tries to explain this difference by a very significant illustration: 'And doubtless this theory, proved by the facts—namely that some incorporeal kinetic force is imparted by the thrower to the body which is thrown during the time the thrower is in contact with the missile—is no more difficult to accept than that certain forces reach the eyes from objects which are seen, as Aristotle thinks. Indeed, we can see from the colours which stain solid bodies exposed to them, that certain forces of an incorporeal form are emitted when the sun's rays pass through a transparent coloured object. . . . It is thus evident that certain forces can reach bodies in an incorporeal way from other bodies' [75].

Philoponus thus regards the emission of light as another kind of impetus, radiated by the luminous source and transferred to the illuminated object. We shall return at a later stage to this very important conception of light which is by no means identical with Aristotle's. It is perhaps worth while at this stage to point out the use made by Philoponus of two different terms for the impetus— kinetic *power* (*dynamis*) and kinetic *force* (*energeia*). His indiscriminate use of both of these terms for a concept having some kinship to our modern 'energy' is a long way from the Aristotelian use of the terms *dynamis* and *energeia* for potentiality and actuality.

3. *Natural motion and weight*

The associated concepts of natural motion and weight also had a long and vacillating history in post-Aristotelian antiquity, to no lesser extent than that of forced motion. The same picture, so characteristic of antiquity, is presented to us: acute discussions of the significance of concepts, references to observations which are sometimes combined with a false interpretation of facts, conclusions often not checked by experiment, and an inconclusive, drawn-out struggle between the conservative adherents of Aristotle and the bold, critical partisans of new ideas and precursors of a new era. Isolated advances of great importance, such as the work of Archimedes on specific weight, did not become starting-points for decisive progress across a broad front; they did not succeed in breaking through a system of reasoning and deduction common to both Aristotelians and anti-Aristotelians, as the experiments of Galileo were to do. But in an age of the pronounced preponderance of great systems of thought with little technological progress, and when newly discovered facts could hardly shake a theory whereas conceptual inconsistencies could be its undoing—in that long era of later antiquity, every criticism of an accepted picture of nature and every argument against it were of great historical significance. They established the continuity of a critical attitude in scientific thought as a heritage for Arabic and medieval science and secured its active and creative share as an integral part of civilization across the long span of time until the advent of the modern era.

Natural motion according to Aristotle's doctrine was bound up with the teleological conception of the striving of a body to reach its natural place. It was quite logical for him to make the assertion that the heavier a body is, the greater its desire to reach that place, and therefore to make the incorrect assumption, that the velocity of a falling body is proportional to its weight.[24] Consequently he was led to explain the acceleration due to gravity—which was generally known at his time and was proved by inference from other facts by his pupil Strato[25]—by an increase in weight of the falling body with diminishing distance from the centre of the earth. The relevant passages in Aristotle's *De caelo* are not always clear,[26] but there is no doubt as to their interpretation by his orthodox followers, as one can learn from their polemics against opposing views. One of these views, which was widely held, was that the higher the position of a body the greater the quantity of air below, through which it must force its way. Thus, acceleration is due not to an increase in weight but to a decrease in the resistance of the medium as the column of air below the falling body becomes shorter. However, there were also views, diametrically opposed to that of Aristotle, held by no lesser personalities than Hipparchus and Ptolemy (about A.D. 150): 'Concerning weight, Hipparchus claims the opposite of Aristotle. He maintains that bodies are heavier the greater their distance [from their natural place]' [204]. And further: 'The mathematician Ptolemy in his book *On Weights* holds the view opposite to that of Aristotle's doctrine and tries also to prove that neither water nor air possesses weight in its own place' [225].

The reply of Alexander of Aphrodisias to Hipparchus, quoted by Simplicius, is in strict conformity with the true Aristotelian spirit: 'If it is according to nature for a heavy body to be below— this being the reason for its moving towards that place—bodies should be completely heavy and assume their proper form in this regard when they are below; and, attaining their perfection through a downward pull, it is reasonable that they should increase in weight the nearer they come to that place. Indeed, it would make no sense to say of bodies that perform their natural motion downward with increasing velocity the greater their distance from above, that they would exhibit this property if becoming less heavy' [205]. In another passage Alexander explicitly refers to Aristotle: 'A better reason, more in accordance with nature, for the acceleration of bodies the nearer they come to their proper places, is their increase in weight or lightness, as Aristotle says. Aristotle would ascribe it to the fact that the nearer a body approaches to its proper place, the purer the form it attains, that is, if it is a heavy body it will become heavier, and if a light one, lighter' [206].

Simplicius himself remarks that Hipparchus' view could lead to absurd consequences in the case of a balance since the side bearing the heavier weight, on being pulled down, would become lighter than that bearing the lighter weight and would thus start to rise.

As nothing has come down to us of the arguments of Hipparchus, we can only guess at his reasoning. First we must remember that Hipparchus had quite another explanation for the acceleration of freely falling bodies and thus had no reason to relate their greater velocity to an increase in weight. On the other hand, Hipparchus, like Aristotle, may have regarded weight as the desire of a heavy body to reach its natural place and from this could easily have drawn the opposite conclusion and said that, having reached that place, the body could lose its natural pull which exists only as long as the body is elsewhere. Moreover, if, as Hipparchus has assumed, natural pull is proportional to the distance from the natural place, a heavy body would reach weightlessness only at the centre of the earth, and the fact that on its surface bodies do have weight would not contradict his assumption. We must further bear in mind that weight, which is a property of the natural pull of gravity and is characterized by 'natural motion' downward, was not necessarily identified with weight as measured by comparison with a standard on a simple balance. What is involved in the latter case and which today is associated with the concept of mass (spring balances were not known in antiquity) was also expressed in the Greek language by a special technical term of its own, which was different from that for natural pull. We will return to this important point when discussing the conceptual analysis of weight in Greek science.

Not until the writings of Simplicius do we find on record a suggestion that experiment should decide between Aristotle and Hipparchus. This need not astonish us, for the usual balance based on the principle of the lever would be of no use in this case. We do not know whether Simplicius had in mind some primitive form of spring balance or whether he thought that the difference in weight would be great enough to be felt by the muscles of a stretched arm. Simplicius expresses doubt as to the truth of Aristotle's assumption: 'Personally I think it worth while to investigate how the acceleration of bodies as they approach their natural places—a fact admitted by everyone—is to be explained. If an increase in weight or lightness occurs, it will follow that an object which has been weighed in the air (the weigher leaning out from a high tower or tree or cliff) will be found heavier when it is weighed on the level of the earth below; this would seem to be untrue, unless it is claimed that the difference between the two cases is imperceptible' [207].

Ptolemy with his anti-Aristotelian view that water and air in their own places have no weight, seems to have followed in the footsteps of Hipparchus. Water is not an 'absolutely' heavy element like earth or earth-like substances, so that its natural place is above earth, within convenient reach for an experimental proof. Indeed, Ptolemy believes that he has evidence for his view, as Simplicius relates (however, without being convinced): 'That water has no weight [in its own place] he shows from the fact that divers do not feel the weight of the water above them, even those who dive to a considerable depth. Against this one may argue that the continuity of the water supporting the diver from above, below and both sides has the effect that he does not feel the weight. . . . However, if the water pressed separately from above, its weight would probably be felt' [226].

In the case of the air, there was conflicting evidence. Aristotle believed that an inflated skin weighs more than an empty one and explains this fact in the fourth chapter of the fourth book of his *De caelo* by stating that only earth is absolutely heavy and fire absolutely light, whereas air and water can be both heavy and light, according to the circumstances.[27] In spite of Archimedes, this explanation held the field until modern times. Simplicius tells of a series of experiments with inflated skins in which the varying results reflect the poor performance of the balances used; these did not permit a conclusive decision to be made, since somewhat higher precision is required. We learn that Ptolemy's results contradicted those of Aristotle—he found the inflated skin lighter than the empty one. One of Simplicius' predecessors got the same result, whereas Simplicius himself, experimenting 'with the greatest possible precision', as he says, found no difference in weight between the inflated and uninflated skin. He thinks that this result agrees best with Ptolemy's assumption that air has no weight in its natural place. 'This is reasonable,' he adds, 'for if natural pull is a striving towards the proper place, objects which are there should not strive towards it nor pull in that direction, since they are already there. He who has had his fill does not reach for food' [227].

It is interesting to see how Simplicius puts greater trust in Ptolemy's theoretical considerations than in his experimental results. However, he says that Ptolemy's results do not disprove his theory, and then develops an idea derived from the confusing Aristotelian doctrine of heaviness and lightness. 'Even should the inflated skin weigh less than the uninflated, it still does not follow, I believe, that elements experience pull in their natural place, but rather proves a lack of pull. For the air in the skin keeps the skin in the natural place of the air, whereas the collapsed skin has an earth-like constitution, and, in the absence of air, the downward pull on it increases. It is like a piece of wood floating on water and partly sticking out into the natural place of air, because through the air enclosed in it, its earth-like part also settles in the place of air' [228]. This passage goes on to show how theoretical arguments weighed much more with the ancients than isolated experimental facts. It is an example of the irony of history that Simplicius did not accept Aristotle's results because he was biased against them by Aristotelian ideas: 'That air should have weight in its own place and that therefore the inflated skin should weigh more would be still more paradoxical. It would mean that air experiences a downward pull from its own place by a natural motion, which, I believe, cannot make sense according to Aristotle's doctrine. But as Aristotle says that the inflated skin weighs more than the empty one, and as it is not easy to ignore the judgement of such a precise man, one can perhaps, with all due respect to him, make the following suggestion: the air which fills the skin is humid, coming generally from the human mouth, and this adds some little weight in the continuous process of inflation' [229].

All of these arguments and artificial constructions, this mixture of apparently logical deliberation and fanciful reasoning based on preconceived ideas and inconclusive evidence, could be seen as an amusing illustration of the scientific frame of mind in later antiquity if it did not give rise to the tantalizing question: How could this state of confusion persist after Archimedes' discovery of specific weight and its determination? This question would not be so vexing if one could say that the theory of Archimedes, still discussed by Vitruvius[28] at the beginning of the Christian era, had been completely forgotten in the time of Ptolemy c. A.D. 150. The whole situation is made so much more disturbing by the fact that the theory and practice of specific weights was very much alive in a time very close to that of Simplicius. A didactic poem 'On Weights' in Latin hexameters, probably written in the early fifth century A.D.,[29] gives a detailed description of the immersion method of the 'Syracusian Master', emphasizing that waters from different springs and different kinds of wine can have different specific gravities, and even describing the construction and working of a hydrometer for the determination of densities of liquids. This poem, which was apparently intended for the use of Roman physicians in the preparation of their prescriptions, leaves no doubt about the continuity of the Archimedean method, based on his theory of specific gravity, throughout antiquity. It seems hardly credible that this method was not known to Ptolemy or to any of the Aristotelian commentators, and one is confronted with the task of explaining how two so basically different scientific theories as those of Aristotle and Archimedes could exist side by side for such a long period of time.

The main reason for this seems to be that these theories did not then appear as contradictory as they do to us. Archimedes' method came to be considered as a technical device not aiming at an explanation of the 'real' physical causes of the variation of specific gravity. Aristotle's view that compound solids were mixtures of 'absolutely heavy' earth and 'relatively heavy or light' water and air could still be reconciled with the ideas of Archimedes as long as his theory was regarded as a purely phenomenological or pragmatic one. Further, one must not underrate the great fascination of the doctrine of opposites in every field of natural science and medicine throughout antiquity and the Middle Ages until modern times. In at least one field it has survived until today, firmly established on experimental evidence— namely, in electricity with its opposite charges. There was also the conceptual difficulty of comprehending a dimensional quantity such as specific weight which involves the combination of two basic quantities, weight and volume. Aristotle occasionally mentions that 'dense differs from rare in containing a greater quantity in an equal volume' [25], and Simplicius in his commentary on this remarks that, when equal volumes are considered, dense becomes identical with heavy and rare with light. But otherwise Aristotle's casual remark was not pursued or amplified, nor was any scale of varying densities ever considered, to replace the extreme opposites 'dense' and 'rare'.

One other factor must be considered which had a retarding and confusing influence on the development of this chapter in antiquity, namely, Aristotle's classification of the pair of opposites heavy–light among the qualities.[30] Natural pull which characterizes both of these is thus definitely not a quantity, for quantities, according to Aristotle, have no opposites.[31] However, he is not so sure about the character of the opposites dense and rare. He feels that the relative positions of the parts composing a dense or rare body play a role here. This shows that Aristotle may have had an inkling that matter as well as volume enters into the definition of density, but he is never clear on this point, and according to his most authoritative interpreter, Alexander of Aphrodisias, he considered weight to be a quality. A number of later philosophers took strong exception to this view, and this conceptual struggle is again a most revealing chapter in the history of the quantification of physical concepts.

Simplicius in his commentary on Aristotle's *Categories* tells us that Archytas, Plato's contemporary, and after him Athenodorus (first century B.C.) and Ptolemy regarded weight as a quantity. Archytas is quoted as having stated: 'There exist three different kinds of quantity; there is quantity in natural pull, e.g. in the weight of a talent; in extension, e.g. in a length of two cubits; and in multitude, e.g. in the number of ten' [136]. This is a reasonable, if not complete, classification of quantities, defined by independent operations which cannot be reduced to each other, that is, enumeration, use of a yardstick, and use of a balance. Compared with it, Aristotle's classification in his *Categories* is, quite definitely, a retrograde step.[32] He distinguishes between discrete and continuous quantities and lists amongst the first, number and speech, and amongst the latter, lines, surfaces, solids, place and time. Ammonius (fifth century A.D.) commenting on this says: 'Some declare the principal species of quantity to be three: number, volume and force, i.e. natural pull. They say that speech and time are identical with number, that lines, surfaces and solids are all part of a common concept, namely extension, and that place is the same as surface. There are thus three quantities —number, extension and force' [6].

The identification of force and weight is no less remarkable than the insistence on regarding weight as a quantity. Simplicius continues the argument from another angle, and refers to Aristotle's statement that the predications of quantity are 'equal' and 'unequal', whereas those of quality are 'like' and 'unlike'. Simplicius then continues by saying that Alexander followed Aristotle in putting weight not among the quantities, but among the qualities, that is, that heavy objects are not equal or unequal, but like or unlike. Alexander is now quoted verbally: 'One usually says that one object is ten times as white as another, not because a tenth part can be assigned to white, but because this can be done to the surface which is coloured white. White is thus measured only accidentally, and the same applies to heavy since it is the body in which heaviness resides that is measured. Thus, heavy is a quality, not a quantity, for every heavier body is at the same time also larger than the lighter, i.e. it has greater dimensions. But if heavy and light are not [directly] measurable, one cannot predicate them by "equal" and "unequal".' Against what Alexander says at great length in the same context—continues Simplicius—'more recent interpreters, following Archytas, have postulated three kinds of quantity and pointed out that weight is not measured [indirectly] through the body. For often body and weight have opposite relations, as for instance in the case of lead and wool whose corporeal dimensions are different when their weights are equal, and, conversely, whose weights are unequal when their volumes are equal. Generally speaking, in all bodies volumes differ from weights, especially as extension and weight are different species of quantity. The *mine* and the *talent* differ in species from the foot and the cubit, so what is measured by a special quantity is not a quality nor is it contained in any of the other quantities. Alexander assumes a *priori*, I do not know how, that, in principle, only extension is a measurable quantity, and not weight; he therefore concludes that weight is a quality. However, were heavy objects to differ in quality, the difference would express itself in a certain variation of features, as one can see in other things which differ in quality. But weight *qua* weight has the same specific character and varies in the amount of its pull according to the law of tendency towards the centre' [137].

The 'more recent interpreters' mentioned by Simplicius were probably Neo-Pythagoreans and Neo-Platonists such as Iamblichus (about A.D. 300), who is quoted by him in another passage as having emphasized the quantitative character of natural pull and its difference from extension and number in that it can be recognized in the motion of heavy and light bodies.[33] It is interesting to find throughout the literature of antiquity the ambivalent use of heavy and light as opposites in the Aristotelian sense, and as different degrees of heaviness in the sense due to Plato. The latter use appears very clearly in a passage in which Simplicius refers to the kinship of the two terms 'natural pull' and 'weight', the latter in the sense of the standard measure used in the operation of weighing on a balance. 'Heavy and light with regard to weight describe a quantity which is subject to a stronger or lesser natural pull, in the sense of Archytas' definition that natural pull and weight are one and the same species of quantity' [138].

The same diversity of opinions persisted of course also with regard to the conflicting conceptions of weight as quantity or quality. Philiponus, for instance, adhered to the Aristotelian view. But even though he includes weight among the qualities together with colour, heat, cold, etc., he discusses at length some characteristic distinctions between different species of quality, as he sees them. There is no change in the intensity of colour, he says, when its quantity is subdivided or multiplied; different amounts of the same white lead display the same quality of whiteness, just as different amounts of water at the same temperature have the same quality of cold, the whole not being colder than each of its parts.[34]

This does not apply to heaviness, for weights are additive and this property of additivity, as Philoponus remarks, is independent of the surface of contact between the weights. 'When a ten-pound weight is put on another body it does not touch the body to which it is added with its whole bulk, but only partially, that is, with its surface, and yet, it acts with the total force that resides in its entire volume. For the pressure creates the weight as a whole' [70]. Heaviness is thus concluded to be a volume force which acts instantaneously; Philoponus contrasts this with the quality of heat and cold where the immediate environment of two bodies exerts the main influence on their thermal behaviour, with the more remote parts playing only a negligible role.

Shrewd as these observations of Philoponus may be, they cannot be compared in importance to his theory of impetus and to his contribution on another aspect of the problem of weight. He is the first on record in antiquity to polemize against Aristotle's doctrine concerning the velocities of falling bodies and their dependence on the weights of the bodies and the densities of the media through which the falling motion takes place. On this point, a view of long standing already existed—that of the atomists who accepted the vacuum as a given physical fact. In his letter to Herodotus, Epicurus pictures the movement of atoms through empty space as follows: 'When the atoms travel through the void and meet with no resistance, they must move with equal speed. Heavy atoms will not travel more quickly than small, light ones, so long as nothing meets them, nor will small atoms travel more quickly

G   85

than large ones . . .' [49]. Lucretius in the second book of his *De Rerum Natura* amplifies this and attributes the dependence of the velocities of falling bodies on their weight to the different resistance offered to bodies of different weights by the surrounding medium.

This Epicurean view (correct, as we know since the invention of the air pump) was of course diametrically opposed to the Aristotelian, for the simple reason that Aristotle denied the existence of the vacuum. According to his view, a well-defined motion could only take place in a material medium of defined properties. Some of his conclusions (in the eighth chapter of the fourth book of his *Physica*) can be summarized as follows: (1) The velocities of falling bodies (in a given medium) are proportional to their weights. (2) The velocity of a given body in different media is inversely proportional to the densities of these media. (3) If a vacuum could exist, it would be like a medium of zero density, and thus the velocity of a falling body *in vacuo* would be infinite, irrespective of its weight. (4) If one were to assume the existence of a finite velocity in a vacuum, one could construct an absurd case in which a body would fall through a certain medium and through the vacuum with equal velocity.

Philoponus, in his Corollary on the Void (in his commentary on the *Physica*) takes strong exception to Aristotle's theory. He adheres strictly to his own view—which he also expressed in his theory of the impetus and of forced motion—that the medium can only have an impeding influence on the falling body, and therefore that the case of 'pure' motion is realized only in a vacuum. He retains the Aristotelian doctrine that bodies strive to reach their natural place with a velocity proportional to their weight, and thus concludes that the law in its pure form can hold only for bodies falling freely through a void. In another respect, motion through media does not obey the laws presupposed by Aristotle, who 'wrongly holds that the ratio of the times required for the motion through various media is equal to the ratio of the densities of the media' [76]. Philoponus refutes Aristotle by an imaginary experiment. Instead of considering the same body falling through media of different densities, he considers bodies of different weights falling through the same medium. Accepting Aristotle's assumption, one could argue on grounds of plausibility that 'when there is one and the same given medium with

86

bodies of different weights moving through it, the ratio of the time required for the motions would be equal to the inverse ratio of the weights; for instance, if the weight is doubled the motion takes place in half of the time' [77]. But, continues Philoponus, 'this is completely wrong, as can be proved better by the very facts than by theoretical reasoning. For if one lets fall, simultaneously, from the same height, two bodies differing greatly in weight one will find that the ratio of the times of motion is not equal to the ratio of the weights, but that the difference in the times is very small' [78]. From this, Philoponus concludes that 'it is reasonable to assume that when identical bodies move through different media, such as air and water, the ratio of the times of motion through these media is not equal to the ratio of their densities' [79]. It would then follow that if, for instance, the medium is half as dense, the time of motion would not be reduced to half, but could be longer than that. Extrapolating to zero density, Philoponus arrives at the conclusion that Aristotle's view of the whole question of free-falling bodies is not tenable.

In spite of the fact that Philoponus' ideas about the velocity of falling motion *in vacuo* constitute a step backwards when compared with those of the Epicureans, his scientific reasoning shows considerable maturity and is obviously superior to that of his predecessors, including Aristotle against whom he is arguing. This maturity of thought and the carefulness of his reasoning are due, of course, not only to Philoponus' own highly original and intuitive mind; when viewed in the context of history it is seen also as representing the progress made by scientific thought during centuries following the Aristotelian period when both philosophers and scientists, inspired by the advances in mathematics, astronomy, technology and the scientific argumentations of the various philosophical schools had made great progress in the discipline of the mind. It was through the steady accumulation of these achievements that the sophistication and flexibility of thought had developed to a degree that is reflected in the writings of men like Philoponus.

Scientific language, too, had become more articulate in spite of the repetitive and sometimes cumbersome style of later antiquity. There was an increased tendency to illustrate general ideas by concrete examples, a greater emphasis on a more detailed and factual formulation of laws and description of processes, and frequently a transition from a purely geometrical description to one with more physical and technical content.

In connection with this, it is illuminating to compare the accounts given by various authors throughout antiquity of the composition of velocities. The fact that two simultaneous motions of a body, taking place in different directions, add vectorially to a resultant along the diagonal of the parallelogram defined by the two directions, has been known at least since the time of Aristotle. Aristotle himself in his *Physica* comments only generally on the fact that two contrary motions in a straight line or in a circle cancel each other, whereas 'a sideways motion is not the opposite of an upward motion' [17]. This obviously implies the idea that two such motions do not neutralize each other, but result in a motion in another direction. In the pseudo-Aristotelian *Mechan-*

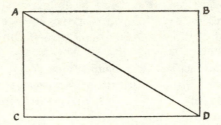

ica, written by an early Peripatetic, we find a geometrical description of two displacements along the sides AB and AC of a rectangle ABCD: 'Let the proportion of the two displacements be as AB to AC, and let A be brought to B, and the line AB brought to CD. . . . The point A will be on the diagonal' [41]. In Hero's *Mechanica*, published perhaps about A.D. 100, a more kinematical description is given: 'Let the point A move with constant velocity along AB and the line AB move with constant velocity along the lines AC and BD, . . . and let the time in which A reaches B equal the time in which AB reaches CD . . .' [58]. A third description which appears in Philoponus' commentary on the above-mentioned passage in Aristotle, is of a dynamical character and illustrates the law by a collision of two bodies: 'Let a rectangle be given and two bodies in motion, one starting downward along one side, the other starting laterally along the other side. If they meet near the earth they will not stop each other but will collide and move with an oblique motion in the direction of the diagonal of the rectangle. Obviously these are not contrary motions (for they do not stop each other), but so to say are sub-contrary because they hinder each other in their original direction but not in their motion as such' [81].

Although it is a long way from Aristotle to Philoponus, there is a tradition of time-honoured terminologies and they tend to persist; Philoponus feels obliged to remain faithful to the Aristotelian term 'contrary motion' and thus invents the term 'sub-contrary', obviously borrowed from logic, for the case of an oblique collision.

4. *Motion of continuous masses*

Still more conspicuous and certainly more far-reaching in its conceptual implications is the development of scientific language in the description of the motion of continuous masses. Propagation of physical action within a continuous medium, i.e. wave motion or the bulk movements of continuous masses such as air or water, require greater power of abstraction for their description than movements of isolated point-like bodies. In this respect, as in many others, the situation was much easier for the atomists. Their two elements of reality were atoms and the void, and it was not difficult for them to explain changes in the density of a gas, for example, by assuming changes in the distance between the atoms which constitute the gas, these being pictured as moving freely in all directions in the void. Aristotle's disciple Strato of Lampsacus (*c.* 300 B.C.) had already realized the great advantage of Democritus' basic assumption for the explanation of phenomena involving the interaction of bodies, and in spite of being the head of the Peripatetic School, he developed a theory which constituted a compromise between the rigid Aristotelian concept of continuity and the atomists' notion of empty space in which all atomic and macroscopic matter is embedded. He postulated a granular structure of matter which allowed for some space between material particles. Simplicius tells us that 'Strato of Lampsacus tries to show that the void divides up the whole body so that the body is not continuous, saying that otherwise light or heat or any other physical influence could not be transmitted through water or air or any other body' [156].

In the centuries after Strato we find this tradition of a corpuscular conception of matter continued by Ctesibius, Philo of

Byzantium and Hero of Alexandria. In Hero's *Pneumatica* the compressibility of air is explained by the assumption, already made by Strato, that there are empty spaces between the particles of air. Hero compares the air to grains of sand that are only in loose contact with each other. Through pressure the volume of air can be reduced, and the withdrawal of the force leads to a return of the particles to their original position. The air thus expands again to its original volume like a sponge or a heap of shavings when set free from compression.[35]

Parallel with this tradition which certainly lasted for several hundred years—it was mentioned by Simplicius—there existed the purely Aristotelian tradition, based on a strictly continuous conception of matter and couched in the terms of Aristotle's categories. Aristotle, discussing the question in his *Physica* (IV, 9), asks how one can, while rejecting the vacuum, explain the change of water into air or of air into water without having recourse to absurd assumptions such as the 'bulging' of the universe or a strict one-to-one correspondence between two simultaneous transitions in opposite directions. Aristotle solves the problem by reminding us that as matter is a pure substratum, it is capable of maintaining its identity while changing from one state to the opposite one; in the same way as it can change colour or go from hot to cold, it can also change its volume. The actual state of an attribute, say hot, implies the potentiality of its opposite, that is of cold, and thus the transition from one to the other is just the transition from the potential to the actual. Whatever holds good for the opposite qualities must hold also for the quantitative concept of bulk which is only an attribute of matter: 'The matter of a body may also remain identical when it becomes greater or smaller in volume. This is manifestly the case, for when water is transformed into air the same matter changes not by any addition but by becoming actually what it was potentially before. The same applies to air when it is transformed into water, the one being a transition from smaller to greater volume, and the other from greater to smaller. Similarly, when a large volume of air contracts into a small one or vice versa, the same matter, having potentially one of the volumes, takes it on actually' [12].

Despite the philosophical elegance of this formulation, it is essentially weak as a physical statement. Can Aristotle make us believe that the transition of a quantity like volume from the potential to the actual has in principle the same character as a change in a quality like colour? What is the mechanism of the former transition which involves changes of extension? He does not tell us anything about the underlying physical process involved in the change from an actually small and potentially large volume into an actually large one, or about the possible influences which such a change in one volume may have on other volumes near it. This is a striking example of the danger involved in the application of philosophical categories or principles to specific scientific problems. They may play an important role as heuristic or guiding principles at a certain stage in the development of a theory, such as the teleological principle in biology or the principle of sufficient reason in physics; they may serve *a posteriori* as a useful framework for the classification of new modes of approach in science, as for instance has occasionally been claimed with regard to potentiality in the field of quantum theory; they may be indispensable where the general formulation of scientific theories is concerned; but philosophical principles are certainly not helpful and, indeed, are often harmful when applied in place of a technical description in the analysis of a specific case.

Aristotle's static conception of the continuum and his failure to grasp the physical implications of different states of density in a given medium barred him from deeper insight into this question and from a more precise formulation of the problem of changing volumes in terms of continuum concepts. In large measure, the way for this was paved by the Stoic doctrine of the dynamic continuum which, together with the progress made through the experiments of Archimedes and later of the 'Pneumatists' like Hero, allowed for a purely phenomenological description of phenomena involving the flow of gaseous or fluid masses. The following passage from Philoponus may serve as an illustration: 'Air can contract and condense and can be reduced from a larger volume to a smaller one, and again be rarefied by expansion. It can condense and contract not only by cooling, but also by thrust and compression' [83]. Later on, emphasizing his strict adherence to the continuum theory, he gives a description which sounds like a paraphrase of Boyle's law: 'One can exhaust a great quantity of air from the narrow mouth of a flask, in such a way that no other air can enter the flask. Obviously no vacuum, either complete or partial, is created in between—this is impossible. What happens is that the remaining air has been rarefied and now occupies the whole space. Now, if it is possible for air to be compressed and to contract, and again to be rarefied, one must realize that, when water evaporates from the flask either by boiling or otherwise, the generated air thrusts that adjacent to it, and this the air next to it until the amount of air condensed through compression clears a volume equal to that occupied now by the generated one. . . . The same happens when air condenses and turns into water. For the air near to the condensed air must, by the force of the vacuum, follow the contracting one, and the air near to that must follow the adjacent, until so much air is rarefied that it will completely fill the space cleared by the condensed air' [84].

This remarkable passage is contained in Philoponus' commentary on Aristotle's explanation of growth in his book *De generatione et corruptione* (I, 5). In this Aristotle tries to explain how in organic growth the form of an organism can be preserved despite the continuous change of matter through metabolism. On this point, too, Philoponus is much more articulate in the physical illustrations by which he tries to throw into relief the possibility of a preservation of *gestalt* of a system in a process of continuous flow. In one of these illustrations, he pictures the shadow of an object thrown on a flowing river. The shape of the shadow remains the same despite the constant change of underlying water as long as the continuity of the flow is preserved. Occasional increase or decrease in the bulk of the human body does not change the human form to any appreciable degree. This, also, can be made clearer by the picture of the flowing river, considering it as a whole as well as thinking of each of its parts separately: 'In our example mentioned before, the river sometimes swells because the inflow of water is greater than the outflow, and sometimes shrinks because the inflow of water is less than the outflow, but the river as a whole remains continuous in itself because the various parts of the water continuously change their places, and the place of the outflowing water is filled up without interruption and without leaving a gap. The same happens in our body where a gentle dissipation of matter is going on everywhere, and if it increases somewhere it is compensated immediately in a continuous way by the influx from all sides. As Hippocrates says: One conflux, one union, all things in sympathy. Thus the form is not affected by the inflow and outflow of matter' [85]. The simile of the flowing river is very aptly chosen by Philoponus because it represents a state of steadiness in a twofold way—by the balance of in- and outflow at every point of the river at various times, and by the conservation of the form of the river as a whole at any given time.

Philoponus' exposition is also of interest in another respect which may be mentioned here in passing. He asks why man is not immortal if his whole body is constantly being renewed, old matter being replaced by new. His reply is that certain parts of the human body retain their substance; as this substance ages, it leads to the general decay of the whole organism.

5. *Perturbations*

We shall return later to the Aristotelian categories of potentiality and actuality in another context. In the preceding section we were concerned only with the fact that in later antiquity the shortcomings of purely philosophical terms were felt when a more precise and detailed explanation of physical processes was required. The problem of the motion of continuous masses served to illustrate how a development initiated by Stoic physics helped to overcome certain conceptual difficulties. The difficulties encountered in later discussions of another set of concepts of central importance in Aristotle's philosophy of nature were of a somewhat different character. Here, too, Stoic ideas as incorporated in Neo-Platonism offered a solution of great significance for the general progress made in the approach to physical phenomena.

As was so often the case in Greek scientific thought, the discussions began with an analysis of terminology. Aristotle distinguishes among the phenomena of nature between happenings which are in accordance with nature and those which are contrary to nature. We have already dealt with a well-known example of this in Aristotelian dynamics, i.e. heavy bodies move downward *in accordance with nature* and they can be moved forcibly in other directions *contrary to nature.* Aristotle almost casually uses the synonym 'natural' for the first case: 'It is natural as well as in accordance with nature' [9]. Themistius takes exception to this: 'What is in accordance with nature is also called natural, but this is a minority of cases. For there are natural things which are not in accordance with nature, such as animals deformed from birth. For these, too, are creations of nature which, however, failed and did not proceed in the usual way' [231]. Simplicius is still more explicit: 'We say that natural things are in accordance with nature if they have the perfection proper to them. But there are some natural things which are not in accordance with nature, although they occur in accordance with the activity of nature, as is the case with animals deformed from birth, and generally with things suffering from some privation. . . . One could call "natural" everything that accompanies or happens to the essence of nature as such, for instance infirmity or illness, but "in accordance with nature" can be applied only to things that prove to be in accordance with the *purpose* of nature. Thus we say that to be healthy is in accordance with nature, but to be ill, though coming to pass as something natural, is contrary to nature' [143].

The question raised by these two commentators has a bearing on the teleological conception of nature. Nature behaves like an artist, with perfection as its aim. But in analogy to occasional failures in the arts, failures in Nature may be assumed to be possible. 'If [says Aristotle] in art attempts that have failed were aimed at a purpose which they did not attain, we may assume the same to be so in Nature, namely, that monstrosities are similar failures of purpose in Nature' [11]. Among such 'failures of purpose' one has to count not only calves with two heads or a child with six fingers but also every possible deviation from the norm, including illness.

Philoponus takes issue with his colleagues on their view that a deviation from the normal pattern of nature could be called 'natural'. In his opinion only things in accordance with nature can be labelled natural, and such phenomena as illness or monstrosities have to be regarded in a wider framework as parts of a whole, in order to be considered natural. Philoponus' view is derived from the Stoic idea that if something goes wrong, the event or object in question must be seen as a partial phenomenon embedded in a wider system. In the frame of this wider system, taken as a totality, the wrong is compensated in some way and the harmony of the whole is restored. 'There are many obstacles and impediments to partial entities and movements, but none for the whole' [43], as Chrysippus has put it. In a qualitative way, this is reminiscent of the law of classical mechanics that, in a closed system, the sum of all internal forces vanishes.

His train of thought leads Philoponus to a somewhat different aspect of the physical situation. When something 'contrary to nature' happens to a physical object, one has to regard it as a perturbation caused by outside factors. The intervention of these factors, taken together with the resulting perturbation, restores the phenomenon as a 'natural' one, as something happening in accordance with nature. The notion of the perturbation of a system is an eminently physical one, and in classical mechanics it has found its application and mathematical expression in the so-called perturbation theory. It is worth while to quote Philoponus' rather elaborate exposition in which he combines advanced ideas and a belief in astrology and the prevalence of harmony. This, together with his involved style, is highly characteristic of his time: 'Perhaps there are no things which are contrary to nature in an absolute sense, but one has to distinguish between things of a partial character which are not natural and are contrary to nature, and things which are a whole and are natural as well as in accordance with nature. . . . A deformity from birth is not natural for man nor is it according to nature. But taking Nature as a whole in which nothing is contrary to nature (because there is no evil in the whole), a deformity from birth is natural and according to nature. For it happens when total nature transforms the underlying substance and makes it unfit to receive the proper form of partial nature. What I mean is this: If the environment is disposed in a certain way by the revolution of the heavens and does something to the substance of man at his birth so that he becomes unfit to receive the form normally put into him by Nature, his nature will not attain its purpose because of the unfitness of his substance, and another form will emerge, contrary to nature with respect to its partial character, but in accordance with nature and natural as seen within the totality of Nature. It will not be contrary to nature with regard to this totality, because corruption is not contrary to nature if indeed generation is in accordance with nature. This must be the case, because generation of one part means destruction of another.

'I will give you an illustration that will explain what happens with things contrary to nature: Suppose that a lyre player tunes his instrument according to one of the musical scales and is then ready to begin his music. Suppose, however, that someone else loosens the tension of some of the strings or all of them, or rather let us assume for the sake of this illustration that the strings are affected by the state of humidity of the environment and thus get out of tune. Now the player's fingers move the strings so that a perfect melody would result if the strings were still properly tuned; when the player strikes the lyre thus, the substance of the strings does not perform the melody that he had in mind, but instead an unmusical, distorted and indefinite sound is produced. The same happens with organic nature. When the substance underlying the human form, or that of another being, becomes unfit through the constellation of the revolving heavens, it comes to pass that its actualization fails. And just as we do not say that the sound of the untuned lyre is artistic or in accordance with art, although an artist produces it, neither do we say in the case of organic nature that it was a natural event, because it did not happen according to the well-defined laws of nature. However, with regard to the nature of the whole we do say that it was a natural event, for it is in accordance with the nature of the whole that it destroys one thing when it creates another' [67].

It is a long way from Aristotle's straightforward classification of things in accordance with or contrary to nature to Philoponus' intricate reasoning. This change in view points also to another instance which can be mentioned only briefly here. Philoponus criticizes Aristotle's finalistic comparison of the ways of nature with those of art. For Aristotle this comparison works both ways: 'If a house for instance were a thing made by nature, it would have been made in the same way as it is now by art, and if things made by nature were made also by art, they would arrive at the same form as they have by nature' [10]. Philoponus' approach to the problem is based on the conception that an artist can intentionally create monstrosities, conceived as works of art and 'in accordance with art' whereas nature, whose purpose is always perfection, can only create them as freaks and never in accordance with its intentions.[36] Philoponus' introduction of the concept of perturbation in his analogy of the lyre and his explanation of the influence of climatic changes on its strings amounts to a relativization of classical Aristotelian ideas that were always held as absolute. The terms 'natural', 'according to nature' and 'contrary to nature' make sense only with regard to a system of reference, and if the system is a partial one, the correct terminology might be the reverse of that for a system which is regarded as a whole. The problem of the whole and its parts had been put on a more physical basis by the deterministic conception of the Stoics, but it was always one of the questions that occupied the Greek mind throughout antiquity. Long before the discussion of the interaction between partial systems which finally led to the idea of perturbation, the old question was raised again and again with regard to an organism or to structural entities in general: whether the whole is more than the sum of its parts, and whether the parts can be regarded as existing separately from the whole.

Aristotle, in passing, asks this question in his *Physica* (I, 2), and Simplicius in his commentary on the passage tells us of answers given to it over a span of 600 years, ranging from those of Eudemus, who was Aristotle's disciple, to those of Porphyry, the Neo-Platonist.[37] It is typical of the dialectical mind of the Greeks that the problem was given a paradoxical turn and that it was proved that the whole is both equal to and different from its parts. The proof of the first alternative was that the whole never contains more parts than those which constitute it, and the proof of the second was seen in the fact that the whole represents the parts in a certain order which did not exist when they were separate. A great number of illustrations were analysed in the course of this discussion which can be regarded as a chapter in the history of *gestalt*-theory. The most convincing of these is that of Philoponus who again uses the simile of the lyre that has served so successfully in various contexts, since the time of Heracleitus: 'One must see in what way the whole is the same as its parts, and in what way different. It is different in so far as it results from the synthesis and completion of all of them, but on the other hand, it is the same as its parts as nothing enters the existence of the whole in addition to its parts. What I mean is this: if every string of a lyre were tightened properly, for instance as required for the Lydian scale, and then all were struck separately but not united into one chord, then each string will obviously produce the proper sound. However, the harmony following the union of all strings and accomplished by the combined sounds is different from the former. For the confluence of all creates a form

97

which was not in the parts of the broken chord. Thus the whole-ness of the harmony is different when the strings sound together (even when there is a spatial distance between them), and when they sound separately; it is the same in so far as no tone is added to the partial tones which combine to accomplish the form of harmony' [63].

IV

MODES OF PHYSICAL ACTION

1. *Local action and action at a distance*

IN the last chapter we have occasionally touched upon the difference in the approach to the explanation of phenomena between the corpuscular school and the strict continuists. Each approach was conducive in its own fashion to the progress of physical thought; both contributed to the development of scientific language, and both led to further conceptual differentiations. Some examples which pertain especially to the atomists have been given in the second chapter. This chapter will be largely devoted to a discussion of those ideas concerning the modes of physical action in general that arose in later antiquity.

Aristotle, the continuist, continually emphasizes the contiguity of the acting body and the body acted upon. 'There are four ways of being moved by an external agent, namely by pulling or pushing, by carrying or spinning' [16]. The Stoics who saw all physical action effected and transmitted by the all-pervading *pneuma*, regarded tension as the essential cause of motion within the cosmic continuum. *Pneuma* and air were highly active media, susceptible to stress and capable of propagating impulses and waves. The Stoics thus developed the first full-fledged doctrine of local action which flourished, with modifications and variations, and on varying levels of scientific conception, until modern times.

A quotation from Philoponus may serve to illustrate how the tradition of this doctrine was kept alive until very late in antiquity, and it will prove at the same time the extent to which careful

99

observation of more subtle physical phenomena had developed. In this passage (taken from his commentary on Aristotle's theory of sound), Philoponus actually describes the phenomenon of resonance in detail for the first time. Besides the Stoic notion of the propagation of waves, Philoponus also presupposes the Aristotelian theory of metals from his *Meteorologica* (III, 6), where he assumes that metals contain air because they are produced by the enclosure of 'vaporous exhalations'.[38] Philoponus begins by pointing out that a bell or gong resounds for some time after having been struck only if it is suspended from a thin cord: 'If the bell is held by the hand so that it cannot vibrate after being struck, the sound will not last. The reason is that if the bell is not in motion it cannot move the enclosed air. But why? If I shake it with my hand, or if I strike it with a piece of wood, will the enclosed air not be moved? Why does no other substance produce a sound similar to that of brass or other metal when it is struck? One must conclude that the material struck must have a certain fitness. Probably silver or brass or similar metals resound because they are interspersed with much air which is the cause of sounds. True, there is also air in wood, but it is mixed with earth, and therefore it is soundless, or sounds only a little. Brass and similar substances have an admixture of water that has itself a sound-conducting property' [99].

All this is said in the true spirit of the Aristotelian *Meteorologica* to which can be traced the beginnings of alchemical conceptions. However, Philoponus continues as follows: 'An indication of this can be adduced from wine-cups. If we pass a wet finger round the rim, a sound is created by the air squeezed out by the finger, which air is ejected into the cavity of the cup, producing the sound by striking against the walls. Experimental evidence for it can be brought in the following way: If one fills a cup with water one can see how ripples are produced in the water when the finger moves round the rim. Obviously the air ejected by the pressure of the finger creates the motion of the water. If the cup is completely filled with water, many drops continually splash over the rim. This all happens if the cup is held by its stem, but if the cup itself is held in the hand, no sound is produced at all, when the finger moves round. For the body struck must vibrate softly, so that the air is not pushed out but is emitted continuously into the inner part, striking the walls of the cup, and being reflected towards all of its parts, because of its roundness and concavity; it thus lasts on and is not scattered, and so produces a strong sound' [99].

It is curious that Philoponus here adheres to the *pneuma* conception so rigidly. It does not occur to him that the vibrations of the cup are directly transferred to the water; instead he assumes that the air enclosed in the metal acts as an intermediate agent. On the other hand it cannot be denied that in this special case the Stoic idea of the all-pervading *pneuma* and the Aristotelian notion of the nature of metals mutually support each other.

One can hardly imagine an example more fitting to illustrate the Stoic doctrine of continuity than musical resonance. Still more evident than the case of the metal cup and the water ripples is of course that of the resonance of two consonant strings or of other musical instruments where no contiguity exists between oscillator and resonator. Here the classical Stoic term of *sympathy* or affinity was taken as a most appropriate description of the situation. It is given for instance by Theon of Smyrna (*c*. A.D. 100) as follows: 'Strings are in resonance with each other—if on a string instrument one of them is struck, then the other, by some kinship and sympathy, sounds in accord' [234]. There is no simile in the extant Stoic literature giving a simple illustration of that force which bridges distance and forms a certain union between kindred structures. In later writings there is a passage by Philoponus where forces of sympathy are compared to those which hold a rope together through the twisting and interweaving of its many strands. This intertwining creates a kind of long-range interconnection between elements which are not immediate neighbours.

The passage is significant in many respects. It comments on Aristotle's criticism of Plato's view that the soul moves the body. 'He thinks that the soul moves the body by moving itself, by being intertwined with it' [36]. The verb used here by Aristotle suggests the action of the interconnecting of objects like that of being interwoven or entangled or twisted together. The picture used by the early Stoics for this close connection was 'total mixture' which does not result in fusion, i.e. in a loss of the individual properties of the components, but on the other hand is more than a mechanical, mosaic-like composition. Philoponus introduces a new simile: 'What does he mean by the intertwining of the soul with the corporeal part of the cosmos? I see it as

H   101

follows: there are three kinds of combination—by juxtaposition, as in the case of stones forming a house; by mixture, as with wine and water; and by intertwining, as in ropes. Juxtaposition lacks sympathy, for there is no mutual affinity between the adjacent parts; mixture leads to a fusion of the components; but combination by intertwining lies between the other two. It has neither the lack of affinity associated with a juxtaposition nor the complete commingling of a mixture. The intertwined components are in contact with each other along larger portions. He probably wants to illustrate the bond between soul and body and therefore calls it intertwining and not mixing or mingling. For the soul neither fuses with the body into one substance as in the case of fusion, e.g. in drugs and chemicals, nor is it completely lacking in affinity and relationship to it as in the case of supramundane powers' [94]. Again, a strange element of mysticism unexpectedly enters a sober physical illustration. 'Supramundane powers' is a specifically Neo-Platonic term for a level of reality which lies outside the cosmos and thus cannot be in a state of sympathy with cosmic entities.[39] This notion was as familiar to Philoponus as the alchemical terms for drugs and chemical reagents or the strictly physical and very aptly chosen picture of two intertwined strands of rope.

The first signs of a new approach to the nature of physical action can be seen in later antiquity. This approach, which was diametrically opposed to the still flourishing Stoic tradition of contiguous action, and which also differed from atomism, was the belief in action at a distance; although this began to grow in Neo-Platonic circles, it did not develop into a full-fledged doctrine as it did 1400 years later in the wake of Newton's mathematical theory of gravitation. The rival theories to Stoic continuism were not based in any way on the principle of action at a distance. Although they did not hold that action was propagated by the continuous spreading of a state, they believed in its propagation in the form of emanations, i.e. the streaming of substances or of corpuscles, either atoms or larger particles, that transfer their forces by pushing (as in the case of colliding atoms) or by pulling (as in the case of iron dragged by its emanations towards a magnet). The conception of action at a distance without any intervention whatsoever of an intermediate agent was thus clearly opposed to both continuous and atomistic ideas. That it was even more antagonistic to the Stoic doctrine seems plausible on the grounds of its origins which are decidedly non-scientific and have their roots in the mystical strain of Neo-Platonism. There existed a belief in occult forces and supramundane powers that could exert their influence immediately in a way inexplicable by physical mechanisms; it was believed that certain persons in certain circumstances could reveal magic faculties or could have mystical experiences, sometimes in the presence of others but not shared by them, and that some individuals were gifted with the power of communication with remote or deceased persons. All of these beliefs of course belonged to a time-honoured tradition and were not invented by Neo-Platonism, but they were adopted by leading Neo-Platonists and included in their system in a suitably digested form.

The common presupposition of these beliefs was the incorporeal character of these mystical and magical experiences, their independence of time and space, and their indifference to all obstacles of a physical nature. There could have been nothing more obviously and outspokenly antagonistic to the Stoic ideas of strict determinism, of the continuous and corporeal propagation of physical action and of a cosmic picture describable in purely spatio-temporal terms. The following quotation from Iamblichus (c. A.D. 300), the disciple of Porphyry and a prominent representative of the more mystical trend in Neo-Platonism, is taken from one of Simplicius' commentaries: 'One need not share the view of the Stoics, with whom we will continue to differ, that action takes place by contact and touch. It is much more correct to say that not everything acts by contact and touch, but that action happens according to the appropriateness of the active part with regard to the passive one, and further, that many things are active without any perceptible contact, as we all certainly know. Even in cases where action apparently needs close proximity, the contact is only accidental, because the things participating in the process of acting and being acted upon must be somewhere in space. . . . In those cases where distance between bodies is not a hindrance to acting and being acted upon and to receiving the activity of the active part, this takes place immediately and without impediment. For instance the strings of a lyre resound in spite of their being distant from each other, and naphtha is inflammable at a distance from fire. On the other hand, many things which are in contact do not act at all, such as plaster or some medical drug when put on a stone' [139].

The last sentence is of special significance as it reveals a very sound attitude which is naturally linked with the whole view expressed by Iamblichus: it is often much healthier and more helpful in a scientific investigation to discover that there is no connection between two phenomena, or no influence of one state on another, than it is to discover a new dependence. The constant danger inherent in the Stoic approach, and indeed in that of all beliefs concerning strictly contiguous action is that they must lead to a belief in total interdependence and thus can easily blur all judgement and appraisal of possible differentiations and variations of influence. 'Everything is connected with everything' may lead nowhere, but 'this group of things is not influenced by that phenomenon' can result in the discovery of an important class of invariants. Moreover, the basic attitude of those who adhere to the principle of action at a distance, which makes them look for the 'appropriateness' of one body with respect to another without inquiring into the mechanism of the intervention, can be most helpful in the discovery and formulation of laws of nature, as has been demonstrated by the classical case of Newtonian gravitation where the appropriateness in question is the mutual attraction between two mass particles. These remarks are of course not impaired by the fact that the two examples given by Iamblichus (resonance and inflammability) lend themselves very easily to an explanation by local action, nor are they intended to detract in any way from the great importance of the hypothesis of contiguity which was especially fruitful in antiquity because it helped in the development of models and analogies illustrating modes of contiguous transmission of action. Such analogies have always been important expedients of scientific explanation, and all the more so before the development of mathematical physics.

2. Potentiality and fitness

The history of similes and mechanical analogies in ancient Greece is most instructive in that it shows how the slow but steady progress in technology is reflected in the increasing sophistication of scientific explanation, and how skilfully examples from daily experience were chosen to illustrate more abstract concepts. A conspicuous instance from pre-Socratic times was the revolving wheels of Anaximander by which he explained the revolutions of the sun and stars. Later we have the Epicurean picture of a flock of sheep whose state of rest as a whole is made up of the random motions of its individuals which illustrated how a resting macroscopic body can be composed of moving atoms. As against this, there is the Stoic explanation of local action by the simile of the spider at the centre of its web receiving signals through the vibrations of the threads. A convenient picture of the transmission of movements, or of action in general, from place to place was provided by gear wheels. These probably came into use after Aristotle but before the time of Archimedes. Aristotle in his *Physica* (VII, 2) mentions only four types of motion—pulling, pushing, carrying and spinning. When discussing the hypothesis that the soul moves the body, he says in *De anima* (I, 3) that this could imply that 'soul and body must change position in the same manner'. However, the Peripatetic essay *Mechanica*, apparently written in the early post-Aristotelian period, already mentions the fact that a circle which moves, say, in a clockwise sense, will cause another one contiguous to it to move anti-clockwise. 'Some people arrange that from one movement many circles move simultaneously in contrary directions like the wheels of bronze and iron put into temples' [40]. The globe of Archimedes displayed planetary motions of different velocities with the help of a gearing system, as Cicero indicates in the *Republica*.[40]

Technical devices of this kind must have been fairly familiar to people in late antiquity, as can be seen for instance from the comments on the passage from Aristotle just quoted about the soul as a mover. Philoponus questions Aristotle's contention that soul and body must move in the same manner, if 'manner' is interpreted as meaning direction. 'One may raise the question of whether he is right in saying that the mover will move in the same way as the thing set in motion. For if we look at the mule pulling the wheel we see that the wheel describes a circle but the mule moves in a straight line. Similarly, in the case of the bronze sphere the axis moves in one sense, but the parts driven by it move in various senses, partly in the same sense as the axis, partly in the opposite one' [93]. The bronze sphere alluded to must have been one of those more complicated planetaria which had come into use in the last century B.C. However, what is of importance in our context is the outspoken mechanistic attitude in a man of the sixth century A.D. which allows him to compare the supposedly complex motions of the soul with a sophisticated piece of apparatus.

Many examples such as these could easily be given, but the increasing mechanical-mindedness of later antiquity is perhaps thrown into greater relief by a consideration of the conceptual development of one of the central pillars of Aristotelian thought, the categories of potentiality and actuality. These concepts were of the utmost importance for the physical explanation of change. Water is potentially air, and it becomes actually so, when the quality of cold (one component of the combination cold–humid of which water is made up) is replaced by its opposite warm. The situation is similar with changing colours and other accidental qualities. This explanation of course implies that the body that actualizes a certain state or property must possess a capacity for this actualization even when that state or that property is only a potentiality. In the later post-Aristotelian period an increasing need was felt to express the necessity for such a capacity within the frame of scientific terminology. Potentiality is only a necessary condition for actuality but it need not be a sufficient one. Is every illiterate potentially a man who can read and write? The answer is that he is so only if he possesses the faculty of learning the art of reading and writing, and the same applies to technical processes and natural phenomena. Here, too, certain presuppositions must be fulfilled for changes to be possible. The technical term signifying the sufficient condition for actualization was *epitedeiotes*, meaning fitness, appropriateness or suitability, and it came into use as a definite scientific concept in the second century A.D. Sometimes its opposite *anepitedeiotes* was used to express the impossibility of actualization; a phenomenon A which occurs by necessity has an unfitness for becoming non-A. Thus Alexander says: 'To things that exist or arise by necessity, nature has given no fitness for the opposite, or rather an unfitness for the impossible. For they would have possessed in vain that fitness for a change into the opposite, when in fact being unable to be other than they are' [2].

In his discourse on causes, Sextus Empiricus (c. A.D. 200) observes that fire burns not only because of its nature but also because of the presence of a suitable fuel. Wood, for instance, will only burn when it is dry. 'For just as no burning takes place if the fire is non-existent, so also no burning takes place if the suitability (*epitedeiotes*) of the wood is absent' [134]. This passage demonstrates clearly the significance of 'suitability' or 'fitness' as a sufficient condition for a process to take place. Wood is potentially consumable by fire; however, it is actually so only when it is in a state in which it is fit for burning.

The use of *epitedeiotes* in the sense indicated spread widely especially after the rise of Neo-Platonism. In about the middle of the third century A.D. Plotinus introduced the term in order to illustrate his doctrine of the different degrees of participation in the Intelligible in spite of its presence everywhere as a whole: 'One has to understand the presence as something depending on the fitness of the receiver' [112]. Plotinus' simile is an illuminated medium. The contribution of the medium to the effect of light depends upon whether the medium is transparent or turbid, that is, on the fitness of the medium to receive light. In later Neo-Platonism, *epitedeiotes* described the faculty of human media to have mystical experiences, a faculty only given to those who have the fitness to be influenced by 'psychic' forces, as can be seen for instance in Proclus' teachings, especially in his *Institutio Theologica*.[41] On the other hand, we have seen clearly how Iamblichus was induced by this mystical conception of fitness to emphasize the existence of physical action at a distance as against the orthodox Stoic notion of local action. Similarly we find that late Neo-Platonists in an increasing measure make use of *epitedeiotes* as a physical or technical concept. A few instances may serve as an illustration. In the *De generatione et corruptione* Aristotle discusses the basic requirements for physical action and arrives at the conclusion that 'both the thing acting and the thing acted upon must be alike and identical in kind, but unlike and contrary in species' [31]. There can thus only be reciprocal action of flavour on flavour or of colour on colour. For Aristotle, generation is a process towards the opposite whereby the object acted upon changes into the acting one by assimilation. Philoponus commenting on that passage remarks that these processes require the fitness of the active partner to accomplish this assimilation: 'The density of matter can often prevent a change. Thus the black of the ink of a cuttle-fish will often overpower the white of milk, but never will this be done by the black of ebony. The change into the opposite requires matter to be fit to act and to be acted upon' [86].

Sound is another case where fitness of the material is required, as we have seen already in Philoponus' commentary on a passage in Aristotle's *De anima* (see quotation [99] on p. 100). This passage reads as follows: 'Sound is to be taken in two senses—actual and potential sound. For we say of certain things that they have no sound, such as a sponge or wool, and of others that they have, such as bronze and all things which are solid and smooth' [37]. In this context the terms actual and potential make sense only if Aristotle had in mind the same distinction which Philoponus expressed more clearly by 'fitness'—that certain materials like bronze have the fitness to produce sound when struck and thus are potential as well as actual sources of sound.

Fitness played a further very important role in the discussions on the nature of the soul in the light of Plato's doctrine and Aristotle's criticism. The simile of the lyre had already been mentioned in the *Phaedo*, but Philoponus returns to it equipped with a new terminology: 'In the same way as the man tuning the strings of the lyre makes them fit to receive the form of harmony —for the strings themselves are not harmonic, but their harmony is adjusted from outside by a craftsman—so it is with the temper of the bodies of animals, for they are adjusted from outside by Creation through the fitness of the temper' [64]. The specific physical organization of the body, resulting in the appropriate blending of all its essential ingredients and called here the 'fitness of the temper' also plays a role in Philoponus' comment on a well-known passage in Aristotle's *De anima*. In this Aristotle refutes the notion of the soul as the mover of the body by the following argument: 'If thus the locomotion of the soul were possible, it would also be possible for the soul that had left the body to enter it again, and upon this would follow the possibility of the resurrection of animals which are dead' [35]. The gist of Philoponus' remarks is that Aristotle errs in assuming that, after death, the body is fit to be moved again by the re-entering soul. The soul is a source of energy that keeps the body moving as long as it is in the proper condition to be worked on, to wit—as long as it has the mechanical fitness which, however, it loses when death occurs: 'Some people claim that the soul moves the body so to say by a mechanical device, as if the body were pushed by the motion of the soul, as when children in their play make small, very thin hollow balls of wax and enclose in them some blue-bottles or beetles so that, when these move, the ball is set in motion. Or like an animal which is enclosed in a cage moving the cage by its own pushing movements. . . . But if the soul moves the body and pushes it so to say mechanically, Aristotle says that it could leave the body and then re-enter and move it again, and thus cause the resurrection of dead animals. For what could prevent this if the motion is produced by purely mechanical means? However, one can object to this and say that Aristotle was wrong when he maintained that if the soul moves the body by moving itself, it could re-enter and bring the dead to life. For instance, take a pillar whose action as a lever lifts up a wall or something similar. When the pillar slips the wall collapses and its joints break up and nobody is able to lift it again by applying the lever. One can find other similar examples. The argument is that in such cases mechanical devices alone are not enough, but there has to be a fitness of the object to be lifted. . . . Those explaining motion by some sort of mechanism alone do not attribute to the body a fitness nor any natural capacity for motion. We, however, assume that through the presence of the soul some vital force is implanted in the body, and accordingly its absence leads to a col-lapse of the body. It is therefore probable that the soul cannot re-enter the body after the removal of that fitness which had been implanted in the body at the beginning. . . . Similarly a stick pushed against a door cannot move the door when it has not the fitness necessary for being moved but, having this fitness, it will move when pushed by the stick. But it will not do so when fastened by nails or when the hinges are loose. Everything set in motion by something else generally needs a certain specific fit-ness. . . . Democritus, too, among all the others could have said that assuming that the soul moves the body by moving itself, one must presuppose a certain fitness of the body to receive the motions of the soul, for instance such and such an order and posi-tion of such and such atoms. Others again would have to pre-suppose a harmony of certain elements whose dissolution would make it impossible to push the body, exactly as when the shape or quality of the wax is lost, for instance when it becomes soft or undergoes some other change, the animal enclosed in it cannot move it any more' [92].

An interesting aspect of this remarkable passage is the way in which the hypothesis of the soul as a moving mechanism is treated here as a reasonable assumption, anticipating in a way Descartes' doctrine by more than a thousand years.

3. *Actuality and action*

We have discussed fitness which in some contexts replaced and in others complemented the Aristotelian concept of potentiality. Now let us turn to the subsequent history of actuality, *energeia*, whose significance also underwent many modifications. It is well known that Aristotle himself also used *energeia* in the sense of activity, and in Hellenistic times the use of the concept in this meaning became general, especially in biological and medical writings. Galen in his physiological works uses it more specifically in the sense of function. Thus he says in his treatise *On the Natural Faculties* that the natural blending of the four elements in their correct proportions causes the normal functioning (*energeia*) of every organ of the human body.[42] Hero (first century A.D.), in the *Pneumatica*, also speaks of *energeia* in the sense of function when he describes the working of the siphon.[43] In his book on the building of automata, *energeia* signifies the mechanism of an automaton in the context of Hero's remarks that the *energeia* of a static automaton is safer than that of a mobile one.[44] In this more technical meaning of activity, function, force or power, *energeia* and *dynamis* are often almost interchangeable, whereas they remain opposite concepts as the philosophical terms of actuality and potentiality.

It is most illuminating to follow up the transition from the philosophical to the technical usage of *energeia* in the special case of the theory of light. This at the same time reveals the curious picture of the coexistence of two conflicting theories and their final merger through suitable interpretation of the terminology involved. It also provides a brief outline of the history of the concept of light in post-Aristotelian times.[45]

Aristotle's exposition of the nature of light in the *De anima* centres round the basic categories of potentiality and actuality as applied to the concepts of transparency and colour. According to him light is the state of actual transparency in a potentially transparent medium and thus represents the necessary condition for vision. The sufficient condition is fulfilled if there exists in the actually transparent medium a potentially coloured body which then becomes actually coloured and produces vision. Light is therefore incorporeal and a state whose emergence and disappearance are instantaneous and which has nothing to do with movement and in particular not with locomotion. When Aristotle speaks of movement in connection with light and colour he generally has in mind the transition from potentiality to actuality, i.e. the realization of a definite state.

This conception clearly separates the Peripatetic doctrine from the three other main theories, all of which were formulated before Aristotle and which can all be characterized as emanation theories. First there is Empedocles' hypothesis according to which light is regarded as a 'streaming substance' emitted by the luminous body and propagated with finite velocity. Kindred to this is the theory of the atomists who make the assumption that 'visual rays' are emitted by the eyes of the observer towards the object seen with immeasurably great or infinite velocity. Finally there is Plato's combination of both these conceptions in the hypothesis that vision is produced through the coalescence of the rays emitted by the eyes with the light issuing from the object.

A further antithesis existed between the Peripatetic view and geometrical optics whose origins also are probably pre-Aristotelian. Euclid, Archimedes, Hero and Ptolemy developed it into a mathematical discipline which was concerned mainly with the laws of reflection and refraction and the theory of perspective, and which was later expanded by Arab physicists. The antithesis was based on the fact that geometrical optics also started from the assumption that light rays originate in the eyes of the observer. Geometrical optics was thus tied up with the theory of the atomists which assumed the corporeality of the emitted rays and was therefore rejected by the Peripatetics. However, they would not have accepted this theory even if the rays had been regarded as purely mathematical lines, by reason of the incompatibility of Aristotle's continuum conception with one of the basic assumptions of Euclid's optics. Euclid imagined that the cone formed by the visual rays whose apex lies in the eye of the observer is not completely filled with rays, but that two neighbouring rays are separated by a very small but finite angle. The separation of the rays constantly increasing with the distance from the apex allowed for the explanation of the basic fact of perspective, e.g. of the disappearance of objects which are so far away that they fall within the space between neighbouring rays.

Geometrical optics consolidated its position more and more during the Hellenistic period not only through mathematical work but also by practical application, as for instance the construction of different kinds of mirrors or the use of burning-glasses for the production of fire. This development did not remain without influence on the Peripatetic writers of later antiquity. It is interesting to observe how, in spite of their endeavour to adhere to the Aristotelian view, Alexander, Themistius and Simplicius—consciously or unconsciously—made concessions to the geometrical notions which find their expression mainly in a modification of terminology. This tendency is especially pronounced in the change which occurred in the meaning of *kinesis* from that of transition from the potential to the actual to that of locomotion from the luminous object to the eye. Alexander still keeps as near as possible to the orthodox view. There is a well-known passage in the *Meteorologica* where Aristotle in his explanation of the rainbow uses the terminology of 'emitted rays' and talks of 'our vision being reflected from all smooth surfaces' [34]. This expression is regarded by Alexander of Aphrodisias as an embarrassing deviation from Aristotle's original view and as a 'lapsus calami'. He himself (or the author of the second part of *De anima* attributed to him) rejects every interpretation of light as movement and offers an unmistakable simile: 'Qualitative change is motion and occurs in time and by a gradual transition. But this is not the way in which a transparent medium receives light and colour. It does not undergo a change but rather the situation is similar to someone becoming a right-hand neighbour without any motion or action on his part. Such is the turn which a transparent medium takes with regard to light and colours. . . . And just as the right-hand neighbour ceases to be on the right when the man on his left leaves his place, so does light disappear when the illuminating source is removed' [1]. Light is therefore a state characterized by a certain relation and is certainly not an affection or a modification of the medium.

Plotinus whose discourse on light was obviously influenced by Alexander defines light as *energeia* of the luminous body 'in the outward direction'. *Energeia* is thus used here in the sense of 'activity' rather than 'actuality', but Plotinus emphasizes that there is no question of any emission of effluence whatsoever. The direction of the mechanism, from the object towards the eye, was pictured even more clearly by Simplicius who compared the role of the transparent medium to that of a stick transferring the effect of a blow from the hand to a stone. Themistius used a somewhat different analogy: the transmission of the image to the eye is not to be compared to the impression of a signet-ring in wax which is a process taking place only on the surface. The eye sees the object because the coloured image fills the entire intervening medium.

Whereas the shift from the purely relational conception to a more directional one is clearly recognizable, all the writers mentioned adduce identical or similar arguments against the corporeal nature of light. Its speed of propagation is infinite ('as proved by evidence'); it permeates the (corporeal) air which—assuming continuity—could lead to the absurd consequence of total mixture. This latter absurdity would be avoided by assuming a porous structure of the air and other transparent bodies, but then the image would show discontinuities, which again is against the evidence. Further, light is not swept away or blurred by the wind, and it does not rise although it is lighter than fire; a single body has definite movement, either circular or rectilinear, upward or downward, whereas light expands in all directions. The last of these arguments against the corporeality of light together with Themistius' conception of the total penetration of the medium by the image indicate that the Peripatetic commentators had some unclear idea of the propagation of a state which, however, did not crystallize into a wave conception.

A much bolder approach to the problem was made by Philoponus who expressed an original and independent opinion as on other subjects. He first gives an exposition of Aristotle's view and then proceeds by raising the fundamental question of how this view can be compatible with both the basic facts of geometrical optics and the thermic effects of light, the latter, as experience shows, being so conspicuously enhanced by the concentration of light through burning-glasses. Philoponus' main argument which he keeps repeating is that light cannot possibly be a static phenomenon, because if the whole medium was completely filled with the (static) images of the objects, we should be able to see things behind our backs or, more generally, objects which are not in our line of vision.

Philoponus sees only one way out of this difficulty—the interpretation of Aristotle's *energeia* as a kinetic phenomenon proceeding from object to eye, and the application of the laws of geometrical optics to this phenomenon. 'As a solution for all these difficulties we suggest applying to the notion of *energeia* the hypotheses of those who regard light and visual rays as corporeal. In the same way as these people assume that sight and visual rays are emitted in straight lines and reflected from smooth surfaces according to the law of equal angles, we assume that the *energeiai* of the colours and light are emitted in straight lines and are reflected making equal angles. For this reason images appear in mirrors not because our visual rays are projected to the objects but because the *energeiai* of the objects are projected in our direction. And therefore we do not see things behind us, because the *energeiai* move in straight lines towards our eyes. . . . And for the same reason, exactly as in the theory of visual rays, we see things which are not in a straight line with the eye—for instance things behind us or above or below or sidewards—if the mirror is in such a position with regard to us and the object that the *energeiai* falling upon it and reflected according to the law of equal angles are projected towards our eyes. Generally, all that the others say with regard to the visual rays we say precisely with regard to the *energeiai* and thus save the phenomena. For it makes no difference whether straight lines proceed from the eye towards the mirror or whether they are reflected from the mirror towards the eye. Now if this is the common assumption of those people and of Aristotle's theory, except that he makes it with regard to *energeiai* and they with regard to the visual rays, and as innumerable impossibilities result from the hypothesis of visual rays, one must rather prefer Aristotle's hypothesis which saves the phenomena as well as avoiding the absurdities' [95].

The interpretation of Aristotle given by Philoponus in this significant passage amounts to nothing less than a complete rejection of the Peripatetic doctrine. First of all it is obvious that Philoponus accepts the propositions of geometrical optics and, in order to adapt them to his explanation, he states clearly the principle of reversibility of the path of light for the case of reflection when he says that the incident and the reflected ray are interchangeable. Moreover—and this is the salient point—he introduces a radical shift in the usage of the term *energeia* in its application to light. Whereas before him *energeia* had already been used occasionally in this context to denote activity rather than actuality, Philoponus for the first time uses it unambiguously in the sense of an active entity which is emitted from the luminous object and whose movement, reflection, etc., can be described by means of geometrical concepts. True to the Aristotelian spirit he regards this entity as strictly incorporeal, i.e. it does not behave as mechanical systems do—its speed of propagation is infinite, and it penetrates material media which neither influence it nor are influenced by it. However, he states that one can apply to this entity those assumptions made for corporeal entities, i.e. either for material particles or for the hypothetical particles of light of the opposing theories.

We have already quoted a passage in which Philoponus, conceiving of the emission of light as an impetus, compares it with the impetus imparted to a projectile by the action of throwing. Both are incorporeal and both of them originate from a body— the impetus of the projectile from the thrower and light from the luminous or reflecting body. It is obvious that Philoponus, without the introduction of well-defined physical quantities and without having recourse to mathematical symbols and algorithms, could not possibly have gone further than this analogy. It also seems probable that Philoponus must have had some notion, however remote, that propagation of light means propagation of a *state* which, in contradistinction to locomotion of matter, he regarded as incorporeal.

In his criticism of the Peripatetic view, Philoponus, in a sort of positivistic attitude, poses the question of whether there is any experimental proof of the statement that the illuminated air is statically permeated by images of the object. This he denies, offering the following argument: when the sun's rays pass through a coloured glass one looks in vain for the image and the colour of the glass in the air through which the sun beam passes. Thus the medium, to use the Aristotelian terminology, remains completely unaffected by the shape and colour of the object. However, when there is a solid body in the way, it receives an impression of colour and shape of the glass at the place where the beam is stopped. Philoponus stresses that this fact is independent of the nature of the intercepting body and its surface which may be smooth or rough—it can for instance be the eye of an observer. 'It is thus obvious that in fact the *energeiai* of the objects seen travel through the air in a manner which physically does not affect it and so arrive at the sensory organ and there, as in any solid body, impress the colours and shapes of the objects' [98].

Philoponus is apparently unaware of the wave concept, already developed by the Stoics. This is all the more astonishing as he uses Stoic notions for the explanation of the generation of heat by light. If we accept Aristotle's doctrine that the sun itself is not warm, and if the sun's rays are incorporeal and therefore produce no friction in the surrounding medium, how does the air become warm? Here Philoponus offers an explanation which apparently is influenced by conceptions and analogies going back to Cleanthes and Poseidonius: 'My explanation is this: as we know, the soul without being warm itself, generates in the body some vital *energeia* by which the innate heat is set in motion and which puts life into the living body, but with the departure of the soul the innate heat is immediately quenched. Just so, I imagine, the light in the air, originating from the sun, is nothing other than vital *energeia* which sets in motion the innate heat of the air and warms it. Just as the faculty of passion in the soul, without being warm itself, heats the blood around the heart, and just as anxiety, being an incorporeal *energeia* of the soul, creates heat, so it is quite plausible that the sun which is itself not warm stirs up the heat contained in the air through the movement of the sun's vital *energeia*, i.e. the light' [96].

Production of fire by burning-glasses is explained by Philoponus as follows: The rays are concentrated in a small volume of air above the fuel and are thus more effective in stirring up the innate heat of the air. Philoponus explains meteorological facts in the same fashion. The sun, for instance, warms more at midday than in the morning or evening because in the former case the same *energeia* traverses a smaller quantity of air and thus is more powerful in its effect of stirring the innate heat. He discusses finally also the passage in the *Meteorologica* already mentioned and asks why Aristotle there uses the terminology of visual rays. There can be only one reply, he suggests, namely that Aristotle wanted to use a more popular language. 'He used the clearer hypothesis of the visual rays, for it is not easy to understand that *energeiai* are reflected or that the *energeia* of colours travels through the air' [97].

The doctrine of Philoponus concerning the nature of light reveals a picture already familiar to us: a curious mixture of advanced physical ideas and vitalistic conceptions, a blending of physical reasoning and of a recourse to biological illustrations.

4. *The use of philosophical principles*

In this last section of the chapter, we will touch briefly on an aspect of scientific thought which is of general interest in the history of ideas and exhibits some features characteristic of Greek mentality. It concerns the application of philosophical principles of either a metaphysical or an epistemological nature to physical phenomena and their interpretation. It is a well-known aspect of the Greek philosophy of nature that, since the time of the early Presocratics, philosophical principles were used in cosmology and in conjunction with very general physical statements. There was Anaximander and his invocation of the principle of sufficient reason to prove the state of rest of the earth in its symmetrical position in the centre of the universe. There was Democritus and his formulation of the principle of conservation of that which exists: 'Nothing can come into being from that which is not nor pass away into that which is not' [44].

Aristotle occasionally applies a principle which in fact springs from his teleological viewpoint that everything in nature is done with a purpose: 'God and nature do nothing in vain' [22], he says when insisting that clockwise and anti-clockwise circular motion are in principle one and the same and pointing out that, were one of them taken as being contrary to the other, one of the circular motions would exist to no purpose. Aristotle's resort to the same principle is equally doubtful when he deduces from the spherical shape of the stars, i.e. from their lack of organs of motion, that they do not move naturally by themselves: 'Nature makes nothing contrary to reason or in vain; she must therefore have provided immobile objects with the sort of shape which is least adapted to motion' [24]. This kind of argument, completely divorced from any empirical evidence, is no more acceptable to us today than Aristotle's statement in the *Politica* that because nature makes nothing in vain, it follows that she has created all

I   117

the animals for the sake of man and consequently that hunting is to be commended.[46]

One sees in instances of this kind further confirmation of the view rejecting any interference of philosophy with science, the Greeks serving as a warning. On the other hand, there can be hardly anyone today who believes in a science with no presuppositions including, at least implicitly, philosophical assumptions of some sort. The point in question is the manner in which these presuppositions are compared with, or checked against, statements of fact; one cannot deny the tendency of the ancient Greeks to overemphasize deduction and to indulge in generalized assertions. However, it is well known that general assumptions have been of extreme value as guiding or heuristic principles in the advance of scientific knowledge, especially when followed by a careful exploration of experimental data and used within a theoretical framework whose trustworthiness had been tested independently in other cases.

It is therefore of some interest to examine one or two cases, in later antiquity, in which philosophical principles were applied to physical phenomena, after the great progress made in 'sublunar' physics in the Hellenistic period. One example is taken from geometrical optics, a branch of physics where, as was mentioned in brief in the last section, men like Euclid, Archimedes, Hero and Ptolemy had made considerable contributions. Hero in his book on *Catoptrics*, extant only in a Latin translation, gives a proof of the following proposition:[47] of all rays impinging on a mirror and reflected to the same point, those reflected according to the law of equal angles of incidence and reflection travel the shortest distance. Thus, if MN is the plane of the mirror and A and C the

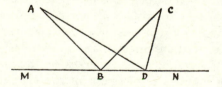

object and the eye respectively, the actual path of the rays, namely ABC, such that AB and BC make equal angles with MN, is shorter than any other path ADC which does not satisfy the law of reflection. Hero adds that the reflection of the rays with equal angles is thus 'in accordance with reason' ('rationabiliter' in the Latin translation). What he meant by this is difficult to decide. However, nearly 500 years later his proof was repeated by Olympiodorus, a contemporary of Philoponus, in his commentary on Aristotle's *Meteorologica*. Olympiodorus does not copy Hero verbatim as can be seen by the slight variations which he introduces into the proof. What interests us here are his introductory remarks, which immediately precede the actual proof: 'It is agreed by everyone that nature produces nothing in vain nor labours in vain. Thus, if we do not concede that reflection takes place with equal angles, it follows that nature does labour in vain with unequal angles, and that the visual rays, instead of reaching the object on a short road appear to arrive there on a longer roundabout way. For one finds that the straight lines which make unequal angles in going from the eyes to the mirror and from there to the object are longer than those which make equal angles' [104].

No matter whether this is Olympiodorus' original idea or whether it is his interpretation of Hero's 'in accordance with reason', the passage quoted in the given context is the first version on record of the celebrated principle of least action, restated in modern times by Maupertuis in the middle of the eighteenth century, and constituting one of the basic tools of physics—today in the final version formulated by Hamilton. Maupertuis' philosophical argumentation has the same outspoken teleological tinge as that of Olympiodorus; so has the reasoning of Fermat who, one hundred years before Maupertuis gave a proof of the same optical law, including it in his 'principle of least time' which he stated as 'Nature always acts by the shortest course.'[48]

The second example concerns an application of the principle of sufficient reason to the theory of the magnet, and is to be found in *Questions and Solutions* which was attributed to Alexander of Aphrodisias but which was probably written some time later, in the third century A.D. The author gives a critical survey of the explanations of magnetic action due to Empedocles, Democritus and Diogenes of Apollonia, and adds his own view. Before coming to the point that concerns us here, a few preliminary remarks are needed on the problem of magnetism in Greek antiquity. The elementary fact that the 'lodestone', i.e. a natural magnet, attracts iron was known since early times, as was the fact of magnetic induction by which the power of attraction can be communicated to a whole chain of pieces of iron attached to the magnet and suspended one from the other—a phenomenon described for instance by Plato.[49] Long before magnetic polarity was known, it was observed that iron (i.e. a magnetized piece of iron) could also be repelled by a magnet. This is stated very clearly by Plutarch: 'Iron often acts as if it were attracted and drawn towards the magnet and often as if rejected and repelled in the opposite direction' [113]. Much later it was discovered that the same magnet has two locally distinct spots exhibiting the polar faculties of attraction and repulsion. This is borne out by a passage in Philoponus: 'By what faculty of the elementary substance does the magnet attract the iron, or is the stone supposed to have the opposite faculty of blowing away and repulsion? And often both phenomena are found in one and the same lump in two different parts of it' [68].

There was a wide variety of theories of magnetic attraction, ranging from mechanistic hypotheses in pre-Socratic times to vitalistic explanations in the later Hellenistic period. The mechanistic theories in some of their variations make use of a wrong principle which one could call the 'inverted jet principle' and which explains attraction by emanations in the direction of the magnet that drag the iron behind them. Another theory, influenced by medical doctrines, is expounded at length by Galen in the first book of his work *On the Natural Faculties*. Plants and animals have the natural faculty of attracting and assimilating what is appropriate and of expelling what is foreign. This faculty is attributed in the spirit of vitalism also to inanimate things like the magnet. Magnetic attraction works on the same principle as e.g. drugs that draw animal poison out of the body.[50] Alexander of Aphrodisias (or the author of the passage on the magnet) develops a similar vitalistic theory of magnetic attraction whereby the simile is a beast attracted to its prey—it is not a mechanical force that acts but an unfulfilled desire.[51]

One of the mechanistic theories—perhaps the first attempt to explain magnetism scientifically—is that of Empedocles, described by Alexander as follows: 'Empedocles says that iron moves towards the magnet because of the effluences from both of them and because of the pores of the magnet which are symmetrical with respect to the effluence of the iron. The effluences of the magnet thrust away the air near the pores of the iron and set into motion the air closing the pores. When this is driven out, the iron immediately follows the stream of its own effluences, for when these effluences move towards the pores of the magnet, because of their being symmetrical and fitting into them, the iron together with its effluences is moved towards the magnet' [3]. Now follows Alexander's critical comment: 'Here the question can be raised—assuming that one accepts the theory of effluences—why does the magnet not follow its own effluences and move towards the iron? For on the basis of this theory there is no reason why the magnet should not be attracted by the iron rather than the iron by the magnet' [4].

Here the invocation of the principle of the lack of a sufficient reason seems perfectly justified. For Empedocles' model exhibits conditions of complete symmetry with regard to magnet and iron and thus there is not sufficient reason to give preference to one of them over the other. In fact, what Alexander claims here on the basis of that principle is the existence of mutual attraction, of simultaneous action and reaction, as is indeed the case, but this was not discovered in antiquity for a simple reason: the magnets used were comparatively large and heavy stones of magnetite, whereas the pieces of iron were small and light; friction was an additional factor that concealed the reciprocity of the attraction. Here, doubtless, we are confronted with a case where the application of a philosophical principle was correct and actually in harmony with the doctrine of the mechanists and atomists who saw in certain conditions of symmetry and likeness a necessary prerequisite for mutual action.

V

CELESTIAL PHYSICS

---

1. *Xenarchus against the aether*

IN turning from sublunar to celestial physics, we must be aware of the extent to which the physics of the heavens more than any other branch of science left its mark on the period before Galileo and Kepler. It was celestial physics which gave scientific and methodical expression to the dichotomy of heaven and earth; when once established through Aristotle's writings, this persistently and successfully survived for nearly two thousand years and was the main obstacle to the inauguration of the new scientific age. The last chapter of this book will deal with the only serious and comprehensive attempt in late antiquity—an attempt not repeated until Galileo—to challenge that fateful division of physics into a terrestrial and a celestial part. In the present chapter we will deal with some earlier anti-Aristotelian developments, but will be occupied mainly with the influence exerted by the regularities of celestial motions on the view of the world in late antiquity and on the vacillations of the various attempts both to describe these motions mathematically and to compare these descriptions with mechanical or other models.

A conspicuous feature of motions within the heavens, apart from their regularity, is their perpetuity, the unceasing diurnal rotation of the uppermost sphere and the proper movements of the sun, the moon and the planets. These perpetual movements constitute a prominent feature which distinguishes the celestial from the terrestrial region where both motion and rest occur and alternate with each other. An apparent discrepancy between this fact and a celebrated definition of Nature given by Aristotle in his *Physica* was the subject of a prolonged discussion: 'All things that exist by nature', Aristotle says at the beginning of the second book, 'seem to have within themselves a principle of movement and rest' [8], and shortly afterwards he repeats that 'nature is the principle and cause of being moved and being at rest'. Are we to conclude then that the heavens which are the primary cause of all that happens on earth are not included in the definition of nature? In other words—have we to regard the heavens as being different from physical bodies? Alexander of Aphrodisias (quoted by Simplicius) rejects this idea and refers to the context in which the definition was given: 'Alexander remarks that the words refer to what Aristotle had mentioned before, namely to animals and plants and the elementary substances, but not to all physical objects. For the heavenly body moving in a circle is physical too, and has within itself a principle of movement, but not of rest, because it moves perpetually' [142]. Simplicius, not content with this explanation, adds one of his own which is rather trivial and outspokenly *ad hoc*: 'One could, however, say that the heavens, too, although they do not change from movement into rest, do rest with regard to their centre, axis and poles, remaining as a whole in their place.'

A much more acute observation is made by Philoponus in his comment on the passage in question. After giving the same *ad hoc* answer as Simplicius he continues: 'Furthermore, as nature looks towards some state of perfection and moves so as to attain it, and once having attained it remains there, and as the heavens are perpetually in that state . . . and do not leave it, they persevere in that state in which they will never cease to be perfect' [66]. Starting from the established idea that perfection with respect to motion means circular motion, Philoponus suggests that the celestial bodies, being in this state of perfection, remain there, continuing for ever to move in circles. In view of the exalted position it held in Greek science, it is not surprising that circular motion should have given rise to the conception of inertia whose essence is just the identification of a state of uniform motion with the state of rest, but with uniformity taking its significance from the physics of Plato and Aristotle, namely regular motion in a circle in contradistinction to the Newtonian concept of motion in a straight line. In the closed world of the Greeks the circle fulfilled the same function as the straight line in Newton's infinite and Euclidian space; each in its respective framework satisfies the condition of geometrical simplicity and guarantees the perpetuity of motion along its path.[52] In another context Philoponus repeats his definition of inertia in a more concise form: 'Rest is also found in all things. For the perpetually moving heavens partake in rest, because the very persistence of perpetual motion is rest' [90]. This should be compared with Euler's interpretation of Newtonian inertia in his *Letters to a German Princess* (74th letter): 'Demeurer dans le même état ne signifie donc autre chose que rester en repos ou conserver le même mouvement.'[53]

The identification of 'remaining in motion' with rest and linked with this the view that the celestial rotation is a kind of inertial motion are certainly of considerable philosophical and semantic interest, but they were not part of the great issue which shook the foundations of the Aristotelian system in antiquity, and which began during the first century B.C. with an assault on the concept of aether, Aristotle's celestial substance. The origins of the attack, which was led by the Peripatetic philosopher and scientist Xenarchus, can be found in inconsistencies and ambiguities in Aristotle's doctrine itself, and it is worth while to recall briefly the relevant points.

The overriding motive for the postulation of the fifth substance, the aether, was to give a physical foundation to the basic assumption of the incorruptibility and stability of heavenly phenomena, in contrast to the perpetual change and fluctuation of sublunar events. Within the strongly classified dynamical system of Aristotle this could only be achieved through the uniqueness attributed to the physical and kinetic properties of the aether. This was easily arranged from the physical point of view by assigning to this substance those qualities of an unchangeable, immutable, sublime material that are lacking in the four other elements which are constantly being destroyed and regenerated. The aether was in fact the embodiment of that highest degree of perfection attainable by matter which can be achieved only in the celestial region. The kinematic aspect presented much greater difficulties to the proof of uniqueness. *Prima facie* it seemed easy enough. The four terrestrial elements were associated with rectilinear motion, i.e. with motion along a line which could be regarded as simple by virtue of its lack of curvature. There was thus reserved for the celestial element a curved and closed line, which is essentially different from the straight one and moreover regarded as simple and perfect because of its constant curvature.

In the second approximation, difficulties began to emerge which Aristotle had to face somehow. The terrestrial elements were paired together as being heavy (earth and water), moving naturally straight down, and light (fire and air) with their natural motion straight up. But circular motion, too, contains a pair of opposites—clockwise and anti-clockwise rotation, and in order to prove the uniqueness of the aether Aristotle had to show that these two opposites were in fact one and the same motion. This he does in the fourth chapter of the first book of his *De caelo* where he defines opposite motions as motions that lead to opposite points, a definition which excludes circular motion as this always leads back to its starting-point if continued in the same direction. In the closed world of Aristotle, the straight line finally leads to a definite place—the natural lowest place for the heavy elements and the highest for the light ones. In these places the respective elements come to their state of rest since they have reached their final goal, their *telos*. The fifth element, on the other hand, is always in its final state, namely in the state of eternal revolution in a circle which is the counterpart of the state of rest of the terrestrial elements. Thus there is another essential difference between linear and circular motion: one is motion *towards* the proper place, the other is motion *in* the proper place. Aristotle's phrase 'motion towards its proper place is for each thing motion towards its proper form' [27] refers to the terrestrial elements, and we have seen in the third chapter how this was interpreted later by orthodox Aristotelians. The circular motion of the aether is the motion of an element that has reached its proper form, and this can be surpassed only by a still more elevated state which is again a state of rest, namely that of the intelligences driving the celestial spheres. However, as these intelligences are not material entities, we need not dwell here upon this continuation of the story which is told in the *Metaphysica*.

So far so good, but Aristotle came up against another difficulty when the situation had to be analysed in the border region between the terrestrial and the celestial spheres. What happens to the air and the fire that rise and reach their natural place at the top of the sublunar sphere? Clearly they cannot remain at rest in the region adjacent to that of the revolving spheres. And indeed, the interpretation of what by Aristotle was thought to be meteorological phenomena—comets, meteors, and possibly the aurora, all subject to 'generation and destruction'—forced him to the conclusion that the uppermost part of the sublunar sphere where hot air and other inflammable material accumulates is carried along with the celestial sphere in a circular motion. He explains all of this at length in his *Meteorologica*.[54] What Aristotle wrote there later became a source of endless confusion and controversy since it is not consistent with what he had laid down in the second chapter of the first book of the *De caelo*: 'a thing can have only one contrary and the contrary of upward is downward' [19], and further, 'that motion which is contrary to nature for one body is according to nature for another, as the motions up and down are according to nature or contrary to nature for fire and earth respectively' [20]. It was just these assumptions that formed an important premise in his conclusion that what moves in a circle must be a fifth element. If now it has to be assumed that fire or something akin to it moves in a circle, the whole hypothesis of the aether becomes artificial and *ad hoc* and there is ample justification in asking whether its introduction is necessary.

Doubts of this kind may have found support in Platonic circles which maintained their tradition that the stars are essentially composed of fire, as is stated for instance in the *Epinomis*: 'What is mainly of fire moves in a perfect order' [106]. In Plato's doctrine of course the perpetuity of these motions was guaranteed by the eternity of the world soul.[55] However, as already mentioned, the first to write against the aether was a Peripatetic. His biography is briefly sketched by his pupil Strabo: 'Xenarchus whose lectures I attended did not spend much of his time at home. He led his life as a teacher in Alexandria, Athens and finally in Rome. He was a friend of Ares and afterwards of the Emperor Augustus and stood in high esteem until his old age. Shortly before his death he lost his eyesight and ended his life in illness' [230].

Xenarchus' book *Against the Fifth Element* is unfortunately lost, and we must reconstruct its contents from the few fragments extant in the form of quotations in Simplicius' commentary on Aristotle's *De caelo*. Xenarchus' attack was apparently aimed at undermining Aristotle's doctrine from several directions, but he does not lay too much stress on the coherence of his own arguments as long as they uncover inconsistencies in Aristotle's system. One of Xenarchus' objections is to the ambiguity of the concept of a simple line. Are the straight line and the circle really the only simple lines? Both have one characteristic in common that each section is congruent to the whole as can easily be seen if one displaces a section along the curve. But there are other curves which have the same property. 'The cylindrical helix is also a simple line, because every part of it is congruent to the whole. Thus, if there exists a simple form in addition to the other two, there should also exist a simple movement in addition to the other two and also, besides the five elementary bodies, there should be another simple body performing that simple motion' [180]. Xenarchus, however, makes it easy for his orthodox Aristotelian opponents to refute his argument by his detailed explanation of a cylindrical helix. It originates, he says, in the combination of two movements, one in a straight line along a side of the cylinder, and the other in a circle along its circumference. Alexander of Aphrodisias (also quoted by Simplicius) now makes the obvious observation that the helix is not simple in the respect that it is generated by two motions differing in character—rectilinear and circular.

Xenarchus finds fault with another basic assumption of Aristotle: 'Circular motion is not that of a simple body according to nature, for in simple bodies that are homogeneous in all their parts, all parts move with the same velocity. But in a circular motion the parts nearer to the centre move slower than those nearer the periphery, because they cover a smaller distance in the same time. This applies also to a [rotating] sphere where the circles nearer to the poles move slower, and the greatest circle moves the fastest of all' [184]. A modern physicist, using the concept of angular velocity instead of linear velocity, could easily have refuted Xenarchus' argument by explaining to him that in circular motion, too, all parts move with one and the same angular velocity, and from this point of view it can be regarded as a simple motion. The reply of the ancient critics of Xenarchus was somewhat different. The gist of their argument is that Aristotle associated the concept of simplicity with points describing lines during their motion. When considering bodies of finite extension one has to consider them as sums of points each of which describes a simple line, and one cannot simply transfer concepts intended for pointlike bodies to extended ones.

Here we witness an interesting discussion that on a much higher level of knowledge could have taken place in modern times. The question was how far one is allowed to go in mathematical simplification when dealing with the description of physical objects and their behaviour. We know today that there is no hard-and-fast rule in this respect, and that in the final analysis, the success usually justifies the means. But the question does make sense and that in fact was what Xenarchus had in mind, as Simplicius tells us: 'Finally Xenarchus finds fault with this procedure in so far as when we inquire about physical things, we make our proofs mathematical and use linear forms in making the cause of simple motions dependent on simple lines' [185].

In another passage Xenarchus mounts his attack on a broader front when he refers to another chapter of Aristotle's dynamics and argues on the ground of a kind of relativism. He says: 'Let us concede that there are two simple lines, the circle and the straight line, and that each of the four elements earth, water, air and fire has essentially one motion according to nature, namely in a straight line. But what prevents some of these elements, or even all of them, from also moving in a circle by nature? For we have not assumed that each has only one motion according to nature; this would obviously be wrong. For each of the intermediate elements has a motion in two directions according to nature: water moves upward away from earth and downward away from fire and air; while air moves downward away from fire and upward away from water' [183]. The final goal Xenarchus had in mind was to prove that fire can also move in a circle 'according to nature', besides being able to move upward, and he very ably makes use of Aristotle's relativistic conception of the intermediate elements water and air of which he said in his *De caelo*: 'Neither of them is absolutely either light or heavy. Both are lighter than earth—for any portion of them rises to the top of it—but heavier than fire—i.e. a portion of them, whatever its size, sinks to the bottom of it' [28].

Xenarchus completes this argument in a rather sophisticated manner by giving his version of the concept of 'contrary' as against that of Aristotle. 'In ethics we say that every virtue has two contraries, e.g. prudence is opposed both to sophistry and silliness, and manliness both to over-boldness and cowardice, and similarly with others. If this is true, there is no necessity to assume that the heavens are composed of some fifth element on the grounds that two motions could not be contrary to one, namely the circular motion of fire and its downward motion to its motion upward. For the motions up and down are contrary to each other like excess to deficiency, whereas rectilinear motion, common to both, is contrary to the circular one like equality to inequality' [188].

Xenarchus also argues that the hypothesis of the aether is superfluous, starting from Aristotle's teleological conception of the rectilinear motion as a motion towards perfection; he elaborates as follows: 'Rectilinear motion is not the motion of any of the four elements in their state according to nature, but only in their state of becoming. But becoming is not a simple state—it is something between being and not being, like motion, which also is between the place to be occupied and that occupied before; becoming is of the same kind as motion, being also some kind of change. In our view, therefore, upward-moving fire is not fire in the proper sense but only in the state of becoming, it is on its way to its proper place above the other elements, and there it remains after it has become fire in the proper sense. . . . It is thus not true that a simple body has a simple motion according to nature, for it has been shown that motion is the attribute of bodies in the state of becoming. But if one has to attribute some motion, and a simple one, to bodies in the state of being, it must be the circular motion. . . . It is thus reasonable to attribute the circular motion to fire in the state of being, and the state of rest to the other three elements' [182].

Xenarchus thus assumes that fire, as soon as it reaches its 'pure' or 'real' state, begins to move in circles, and he has therefore no need for Aristotle's artificial distinction between the circular motion of the 'fiery masses' in the sublunar region and that of the aether in the celestial sphere. When making an appraisal of Xenarchus' anti-Aristotelian exposition it must be borne in mind that we have to rely on a few passages picked out by Simplicius according to what he regarded as the essential points at issue. Logical distinctions and matters of classification always ranked high in the Greek philosophical and scientific tradition, and on a subject like that of the existence of a celestial aether, the character of the discussion would necessarily have been theoretical even had the Greeks been more interested in the experimental aspect of science. We will see in the last chapter that Philoponus, who apparently depended on Xenarchus as far as the passages quoted by Simplicius are concerned, went much further in his polemics against Aristotle and attempted to adduce observational material as evidence for the fiery nature of the celestial bodies. Unfortunately we do not know whether Xenarchus' book also contained some of this material.

The very fact that this book was written is of the utmost significance in the history of ideas. It shows that some of Aristotle's fundamental physical notions were challenged even within the camp of the Peripatetics and in a manner which resounded through the following centuries. Simplicius mentions that Ptolemy and Plotinus, the former in his books *On the Elements* and *Optics*, associated themselves with Xenarchus' view that elements in their proper place either are at rest or move in circles.[56] Plotinus and the later Neo-Platonists of course disregarded the fifth element for two reasons—they were followers of Plato and were the successors of the Stoics, who had denied the existence of the aether and reverted to the four elements.

It still remains to examine the attempts of later commentators to reconcile Aristotle's exposition in his *Meteorologica* concerning the circular motion of fire in the outermost part of the sublunar sphere with his basic assumption about motion in accordance with and contrary to nature. The explanations given by three commentators of the sixth century, Simplicius, Philoponus and Olympiodorus, show an interesting amplification of Aristotelian concepts, clearly bearing the marks of Neo-Platonic and mystical influences. Obviously the two opposite motions 'in accordance with nature' and 'contrary to nature' are not in themselves sufficient to describe the phenomena satisfactorily, and in particular they cannot do justice to the so-called meteorological phenomena, the comets, etc. A third technical term 'above nature', an expression with a strong flavour of Iamblichus or some other Neo-Platonic mystic, was added to the other two. 'The circular motion of fire and of the adjacent air is not natural but is above nature, for they are carried around with the revolution of the heavens' [65], says Philoponus. Simplicius goes into greater

detail and reveals a highly significant train of thought. He compares the motion of that 'fiery region', i.e. the motion of comets
and meteors, with that of the planets. After all, the proper
motions of the planets can hardly be called 'motion in accordance with nature', as they are not regular circular motions, but
neither can they be termed 'contrary to nature', since in the geocentric system as a whole, the planetary spheres are connected to
the outermost sphere of the fixed stars which certainly moves
according to nature.

Let us hear what Simplicius has to say about this: 'As Aristotle
assumes also a circular motion of fire, one must ask if this is in
accordance with or contrary to nature. If it is in accordance with
nature, there would be more than one motion of this kind,
because fire moves upward in accordance with nature. If it is
contrary to nature, there again would exist more than one motion
contrary to another, for fire moves downward contrary to nature.
But perhaps this circular motion of fire is not a proper motion in
accordance with nature because it is carried round with the
sphere of the fixed stars. For that matter, neither is the planetary
motion a motion according to nature as is the motion from east to
west. Nor is the circular motion of fire contrary to nature in the
sense of being opposed to motion which is in accordance with
nature, for it is irregular and not stable and of a different type,
existing through the prevailing force of something mightier.
Perhaps it was for this reason that Aristotle did not call this a
motion contrary to nature but rather a forced motion. Since it is
a helpful force it should not be called 'contrary to nature' but
rather 'above nature' [181]. The trajectory of a missile, which is
produced contrary to nature by the throwing force of the ballistic
machine, is here contrasted by Simplicius with the irregular, but,
as a whole, similar movement of both the comets and meteors
and the planets whose orbits are produced by a 'helpful force'.
This force is helpful because by some mechanism or other—
friction or the intervention of the spheres—it produces a permanent attachment to the sphere of the fixed stars and thus confers on the forced motion a kind of perpetuity. One could draw a
modern parallel and compare this type of forced motion to the
constrained movement in well-defined channels produced by the
restraints of such mechanical devices as rails and hinges.

The comparison with the compelling effect of mechanical

devices must be taken with a grain of salt, because in the age of Neo-Platonism quite different associations of ideas were usually hidden behind technical terms. Plato had attributed a soul to the planets which were thus imbued with a vital impulse moving them in their orbits in the same manner as man is moved by his soul along paths that differ from those 'according to nature'. To some extent, Platonic and Aristotelian elements became mixed up; the origin of that vital impulse was believed to be in the fifth element which was identified with the aether of the Platonic school as described in the *Epinomis*, there being assigned to the intelligences existing on the level intermediate between man and God. This approximately is the background to the following passage from Olympiodorus' commentary on the *Meteorologica*: 'The celestial bodies move the totality of the elements by some contact with them, for it is proved that there exists no void. However, that contact is not of such a kind that a forced motion is generated, because the elements do not move in circles along with the fifth substance contrary to nature, but above nature. Just as a living being, inspired and illuminated by the soul, moves horizontally above nature—while the body has a tendency downward—so the totality of elements moves with the fifth substance above nature' [103]. Similarly Simplicius explains the motion of the planets and the fiery substance above nature by saying that 'along with the other nature of the stronger entity it exhibits a vital motion in accordance with the superior measure of life' [187].

We thus see that the roots of the opposition to Aristotle's fifth substance go back to Plato in a twofold way. It was Plato's conception of the stars being composed of fire, together with inconsistencies in Aristotle's theory, that led Xenarchus to his challenge. Further, it was Plato's belief that each star has a soul which is the driving force in its celestial orbit that, in its Neo-Platonic version, kept the anti-aether bias alive until late antiquity. For Philoponus there was still another reason for this bias— monotheism, which led him to launch his great attack which will be dealt with in the last chapter.

2. *Ptolemy: the struggle for a unitary picture*

We will now look at what is for us the most important aspect of celestial physics, namely astronomy or, still more specifically, the planetary motions. We are not concerned here with the technical details of the development of mathematical astronomy, a chapter which is perhaps the greatest and certainly the most fascinating in Greek science. We will briefly summarize the facts but apart from this will assume that the reader is familiar with the main features of that development which in the five hundred years from Plato to Ptolemy made such astonishing progress. It was a sequence of steps that began with Plato's scientific testament 'to save the phenomena' by a rational geometrical description of the celestial motions, and continued with Eudoxus' theory of concentric spheres, improved by Callippus, and represented as a unified system by Aristotle. Then followed the theories of the epicyclic and eccentric motions, conceived separately but later combined, which led from Apollonius of Perga to Hipparchus, and finally to Ptolemy. These steps consisted in the continual interaction of observation and theory, whereby the desire for more precise agreement led to theoretical modifications and attempts to adapt the geometrical assumptions to the available data. What is of interest to us is the reaction of Greek scientists and philosophers to the increasing complexity of a theoretical system, their attitude towards the problem of model versus reality, their epistemological vacillations when confronted with the question of the 'truth-content' of a theory, and their attempts to reconcile a scientific system with current non-scientific beliefs.

Plato's celebrated challenge 'to save the phenomena' (i.e. to account for them) had a specific meaning: 'By what assumed uniform and ordered motions can the phenomena in relation to the motions of the planets be saved?' [209]. Thus the problem which Plato put to the scientists was to resolve the apparent planetary motions by some kind of 'harmonic analysis' into a sum of uniform circular movements which alone were assumed to represent the divine order of the celestial region, and which, especially for Plato, expressed the eternal system of the world-soul. Thus Plato's disciples were confronted with a geometrical problem, and the solutions offered by Eudoxus and Callippus were obviously restricted to this purely geometrical aspect in a merely descriptive way. The solutions were given separately for each of the planets: each planet was conceived as a point on the surface of a rotating sphere which was the innermost of a series of concentric spheres; these spheres, with the earth as their common centre, were interconnected in such a way that the axis of each was fixed on the surface of the one next to and surrounding it. The directions of these axes and the angular velocities of the spheres were the parameters whose special values 'saved the phenomena' for every planet.

Aristotle, the realist, aimed at a more ambitious goal than a mere geometrical description. For him a description meant the picture of a physical reality whose existence could be derived from some 'true causes', and which formed part of a general system resting on solid scientific foundations. The purely geometrical expedient was thus replaced by a physical model that differed from the pictures of Eudoxus in two essential features. In Aristotle's physics the concentric spheres are material shells consisting of the fifth essence by virtue of which their indestructibility and all the other celestial properties are ensured. Furthermore, the different sets of spheres belonging to each planet were replaced by a unified set that represented one single and ordered geometrical system. This was achieved by the addition of a rather large number of 'counteracting' spheres, interposed between the others whose rotations compensated the undesired effects of the interactions of one set on another and transformed the sphere of the fixed stars together with those of the seven planets into one coherent material system, consisting of 56 spheres.

The intricacies of these 56 aetherial spheres, fitted together like the wheels and cogs of a clock, did not seem to Aristotle to be too high a price to pay for the achievement of a unity not only of a cosmological picture but also of his more general system of the physical world of which the heavens were but one part, albeit the noblest. For now the celestial motions were shown to be one organic and coherent part of the natural movement of the fifth essence, the eternal motion in a circle, as against the finite and perishable non-coherent rectilinear motion of the four terrestrial elements. This was the best that pre-Newtonian physics could achieve in the way of consistency between a set of concepts based on basic principles on the one hand and a well-ordered cosmological picture on the other. The belief in the superiority of a 'physical theory' like that of Aristotle over a mere descriptive device is still reflected in a passage written at a time when Aristotle's system of concentric spheres had already become obsolete. The author of that passage is the astronomer Geminus who lived in the first century B.C. and is known for his *Elements of Astronomy*. A remark made by Geminus in his (lost) commentary on a book by Poseidonius is quoted by Alexander of Aphrodisias; Simplicius copied this which, he says, is inspired by Aristotle and conveys the distinction between an astronomer and a physicist. Whereas the task of the former is only descriptive, the latter turns description into explanation by reaching back for causes and deriving his data from basic considerations. Even Aristarchus' heliocentric theory is mentioned as an astronomer's device to save the apparent irregularities with regard to the sun, in contrast to a physicist's theory which must be based on first principles. A few extracts from Geminus' exposition may be quoted here: 'It is the business of physical inquiry to consider the substance of the heavens and the stars, their force and quality, their coming into being and their destruction; nay, it is in a position even to prove the facts concerning their size, shape and arrangement. Astronomy, on the other hand, does not attempt to speak of anything of that kind, but proves the arrangements of the heavenly bodies by considerations based on the view that the heavens are a real Cosmos. . . . The things of which astronomy alone claims to give an account are established by means of arithmetic and geometry. Now in many cases the astronomer and the physicist will aim at proving the same point . . . but they will not proceed by the same road. The physicist will prove each fact by considerations of essence or substance, of force, of its being better that things should be as they are, or of coming into being and change; the astronomer will prove them by the properties of figures or magnitudes, or by the amount of movement and the time that is appropriate to it. Again, the physicist will, in many cases, reach the cause by looking for a creative force; but the astronomer . . . is not qualified to judge of the cause. . . . He must go to the physicist for his first principles . . .' [144].

In order to appreciate fully the attitude expressed in these lines it is perhaps worth while to draw a parallel from the Newtonian age. A Geminus of the eighteenth or nineteenth century could have contrasted the astronomer Kepler and the physicist Newton who were both concerned with the explanation of the solar system. Newton proceeded along the lines indicated in his preface to the *Principia*. His procedure consisted of this: 'From the phenomena of motions to investigate the forces of nature and then from these forces to demonstrate the other phenomena.' His theory of gravitation 'explained' the planetary motions in so far as this was a special, though extremely important, case of his general dynamics, of a system of physics of the widest possible scope, founded on a conceptual and experimental basis. Kepler's laws, compared with this, were what Geminus called the demonstrations of an astronomer, proofs 'by the properties of figures or magnitudes, or by the amount of movement and the time appropriate to it'. The same situation existed on the lower level of Aristotelian physics which was merely observational and non-experimental and had as yet discovered neither the mathematical tools for the description of physical changes nor even the elementary expressions for physical quantities. The very fact that Aristotle's celestial physics was an integral part of a general conceptual system and fitted into the whole theoretical structure of natural place and natural motion, simple bodies, simple movements, etc., elevated it to the rank of a physical theory that explained and not just described as the pre-Aristotelian models had done.

When Geminus wrote his book, more than two hundred years had already passed since the interpretation of the observational data had forced the astronomers to recognize the inadequacy of the theory of the concentric spheres moving round the earth as their common centre. It was obvious that the distances of the planets from the earth varied appreciably and that other inequalities in the planetary motions had to be accounted for. The two main reasons for these inequalities stemmed, of course, as we know today, from two incorrect assumptions of the geocentric system: that all the motions had to be related to the earth as the body at rest to which everything must be referred (an assumption rejected by Aristarchus, but abandoned finally only by Copernicus), and that the planetary orbits were circles (an assumption replaced by Kepler's ellipses). The development that took place in the four hundred years between 250 B.C. and A.D. 150 is associated with the names of Apollonius of Perga,

Hipparchus and Ptolemy. It is characterized by a continuous and remarkably successful effort to 'save the phenomena' by appropriate geometrical assumptions to which, however, the 'physical' content, the unity of the picture, was sacrificed. Very little was now said about spheres whose very name had suggested material structures, and circles were mentioned almost exclusively in the context of geometrical descriptions. The development began with the eccentric circles, which had already been introduced before the time of Apollonius, whose centres were assumed either as being fixed at a given distance from the earth, or as being movable—the planet moving along the circumference of the eccentric circle while the centre of the eccentric itself described a circular motion. Then the theory of the epicycles was conceived by Apollonius and later re-established and developed by Hipparchus. This theory was preferred by many authors because it preserved the earth as the centre of planetary motion—the planets moving on the circumference of small epicycles whose centres themselves moved along the circumference of the main circle with the earth at its centre.

Many difficulties were involved in the adaptation to a purely descriptive mathematical astronomy which was progressively losing all claims to being a physical theory. Some insight into this problem can be got from a treatise written by Theon of Smyrna at about A.D. 120, only a few decades before Ptolemy published his *Syntaxis Mathematica*. The book was intended as a kind of introduction to the mathematical sciences for the benefit of readers of Plato. In its second part, Theon gives a vivid picture of the state of Greek astronomy in his time and an appraisal of its past history. Here and there one gets a glimmer of nostalgia for the lost paradise of the Aristotelian picture of the world; the following passage is a characteristic example: 'It is not natural that the stars should either move themselves along certain circular or spiral-like lines, and, moreover, in a sense against the rotation of the universe, or that they should be carried along certain circles to which they are fixed and which rotate round their own centres, some in the same sense as the universe, some in the opposite one. For how could such large bodies possibly be tied to incorporeal circles? But there must exist spheres, made of the fifth essence, situated in the depth of the universe and moving there, some higher up, some arranged below them, some larger, some smaller, some hollow and some massive within the hollow ones, to which the planets are fastened in the manner of the fixed stars . . .' [236].

In the first instance one is inclined to judge the degree of reality of a theory by its intuitiveness, by the extent to which it can be demonstrated in the form of a mechanical model or explained in other familiar terms. Since the time of Archimedes such models had actually been built, the first examples being of planetaria which reproduced the system of concentric spheres with increasing technical perfection through the years. Theon, when expounding on Aristotle's reacting spheres, characteristically takes recourse to these artificial models in order to make the reader understand their function and mechanism: 'One had to suppose that between the spheres carrying the planets there were other obviously solid spheres which by their own motions counteracted those of the carrying spheres, rolling them back through their contact. This is similar to the effect of the so-called drums in the artificially constructed spheres that have a certain motion of their own which by the gearing of cogged wheels move in the opposite direction, counteracting the motion of the spheres fixed beneath them. For it is indeed the natural process that all spheres move in the same sense, carried around by the outermost sphere . . .' [237].

Theon's book contains excerpts from a lost work of Adrastus who wrote a commentary on the *Timaeus* probably a few decades before Theon. From these quotations it appears that Adrastus wanted to adopt a model of spheres that differed in two essential features from the Aristotelian. One change consisted in the abandonment of the reacting spheres, which meant the abandonment of the idea of a unitary system of all the planets, and a restriction to the individual treatment of each of them, by separate sets of spheres. The second innovation was the assumption that the epicycle is a circle on a massive sphere rotating between two hollow spheres which could be either concentric or eccentric with respect to the earth. The rotation of these spheres by their contact with the enclosed massive sphere caused the rotation of the latter and thus the epicyclic movement of the planet attached to its surface. Although it is not easy to get a clear picture of Adrastus' idea from the rather corrupted text in Theon's book, it is important to realize that in the first century A.D. attempts were made to return to the construction of some concrete physical picture, as against the pure geometrization of astronomy.

With regard to the geometrical approach, it had been evident since Hipparchus that both the eccentric and the epicyclic methods were equally satisfactory for the description of the phenomena. Theon, however, preferred the latter, not only because it preserved the position of the earth as the centre of the whole system, but also because it made possible the retention of the concept of a sphere to which the planet is fixed. Theon, after describing the eccentric and epicyclic theories, goes on to say: 'For the reasons mentioned it is obvious that the two hypotheses agree with each other but that the epicyclic seems to be more general and universal and closely related to what happens according to nature. The epicycle is the greatest circle on the solid sphere described by the planet when it moves round that sphere. The eccentric circle, on the other hand, is completely remote from things according to nature and is generated rather accidentally. This was also recognized by Hipparchus who preferred the epicyclic theory because it was his own, and who regarded it as more plausible that all celestial bodies should be situated symmetrically with regard to the centre of the universe and should be held together uniformly. However, as sufficient physical data was not available to him, he himself did not realize clearly which of the planetary motions is according to nature and thus a true motion, and which is accidental and only apparent. For he also assumed eccentric circles on which the epicycles move, the planet being on the epicycle' [238].

According to this passage, Hipparchus had already found it necessary to use a combination of the eccentric and epicyclic hypotheses in order to account for the phenomena. From another passage in Theon we learn, on the other hand, that Hipparchus had not yet a definite proof of the equivalence of both assumptions: 'Hipparchus said it would be worth a mathematical investigation to find out the reason why two so different hypotheses lead to the same phenomena' [235]. The equivalence of the eccentric and epicyclic descriptions was finally proved by Ptolemy in his *Syntaxis Mathematica*.

This *magnum opus* of Ptolemy, written in the middle of the second century A.D., and destined to remain the standard work of astronomy until Copernicus, shows amazing progress beyond the already remarkable achievements of Hipparchus. The increasing demand for more precise agreement between the observational data and their harmonic analysis into circular motions led to a still more complex description of the solar and lunar motions as well as of those of the two inner and three outer planets. The phenomena could be saved only by an even bolder extension of the eccentric and epicyclic methods. An increasingly careful treatment of the orbits of each of the seven celestial bodies almost eliminated any prospect of a return to a unitary picture. Greater than ever was the gap between the success of the descriptive method of the astronomers and the failure of their efforts to explain the plurality of the intricate motions by a single physical hypothesis. Ptolemy himself, who was not only a great technician of science, but who had also a strong flair for its philosophical background, was painfully aware of this unsatisfactory situation. Here and there he allows himself to insert a comment on this problem and sometimes he can hardly disguise his disappointment over the lack of a unitary theory of the planetary system. His words have a curiously apologetic air when, in the face of the complex geometrical description, he stresses that the concept of simplicity is relative and cannot be applied equally to terrestrial and celestial phenomena. Here are a few extracts from the thirteenth book of his work: 'Nobody should regard these hypotheses as too difficult, considering the inadequacy of our devices. For no comparison of human things with divine ones can be fitting, nor can arguments be sufficient when they are adduced as evidence on such great matters from the most incongruous examples. Indeed, what can be more dissimilar than things that are eternal and unchanging and things that never remain the same, or things that can be obstructed by anything and things that are never obstructed, even by themselves? There is no other way than to try to adapt as far as possible the simplest hypotheses to the celestial motions and, if one does not succeed with these, to try others that are feasible. However, once every phenomenon is saved in consequence of these hypotheses, why should the complicated motions of the celestial bodies still appear so strange to us? . . . The combination and sequence of all the revolutions may appear clumsy and hard to explain by the models constructed by us; however, in the heavens there is nothing that is obstructed by such a medley. Moreover, one should not judge the very simplicity of the celestial bodies by terrestrial objects which seem to us to be simple because nothing on earth appears with equal simplicity to everyone of us. To those who look at things in this way, nothing that happens in the heavens will appear simple, not even the immutability of the prime motion. For the very fact of always being the same is, when judged from our point of view, not only difficult but nearly impossible to conceive' [127].

Here we see Ptolemy resigned to an attitude which could be called that of a relativist or rather a positivist: we must do our best to give the simplest description possible of complex phenomena which in any case are outside that region where the usual notions of understanding or explanation are applicable. But this was not Ptolemy's entire view of the matter. In a later book, called *Planetary Hypotheses*, he abandons the purely descriptive attitude of a positivist and propounds some ideas about the possible physical structure and causes governing the planetary system. He may already have had these ideas when he wrote his *Syntaxis*, and if so, he refrained from revealing them possibly because of his own doubts as to their substantiality. Two souls dwell together in the breast of many a scientist, and Newton's lifelong speculations on the aether as the cause of gravitation, in spite of his very careful attitude in the *Principia*, presents a famous example. However, the *Planetary Hypotheses* gives us an insight into Ptolemy's attempts not to neglect the role of the physicist—in Geminus' definition—over that of the astronomer. In this book Ptolemy gives a summary of the contents of the *Syntaxis*, and he combines this with several additions and improvements based on new observations. With regard to the physical model, Ptolemy takes up the ideas of Adrastus and of the Platonic writer Dercyllides and develops their hypothesis of the hollow and massive aethereal spheres whose combined rotations make the planets move in their apparent orbits. In doing so he proceeds with great care and, as early as the first chapter of the book, dissociates himself from all the 'classical' models fashioned after the pattern of Aristotle: 'In the present work we undertake only to give a summary exposition of the celestial motions so that they can be more easily understood by us and by those who want to prepare themselves for the construction of instruments—whether one wants, with one's bare hands, to find each of the periods of revolution of the planets, or whether one wants by mechanical methods to combine these motions among themselves and with the motion of the universe. However, this cannot be done by building spheres in the customary fashion. For this method, apart from failing in the presuppositions, presents only the appearances but not what underlies them, and gives only a demonstration of the devices but not of the basic principles. One can do it only in such a way that the order and variety of the motions, together with their irregularities, should, with the aid of uniform and circular motions, catch the eye of everyone who looks at them' [128].

Ptolemy then goes on to say that he will begin by expounding the theory of the celestial motions by means of circles as if they were completely detached from the spheres to which they belong. Only in the second book of the *Planetary Hypotheses* does he proceed to talk about 'the shapes of the corporeal spheres'. The Greek original of the second book is lost and only an Arabic translation is extant, with the exception of a few lines quoted by Simplicius. This quotation gives some idea of the gist of Ptolemy's theory: 'It is thus more correct to let each planet be a source of motion, for this is the power and activity of the planets in their proper places and round their own centre, namely the uniform motion in a circle. The whole motion must thus have its origin in the planet itself which achieves the same thing in the surrounding structures' [133].

The emphasis placed here on considering each planet as an independent source of motion is essentially directed against the unitary model of Aristotle, but it hints also at a Platonic conception regarding each planet individually as imbued with a soul as the source of its motion, as we shall presently see. Before we come to that, a few other passages will make clear the spirit in which Ptolemy conceived his model. He believed in the reality of his spheres, but at the same time his considerations were guided by some principle of the economy of thought that sought to avoid redundant assumptions. He was equally convinced, however, that nature itself always avoids redundancy: 'It is not proper to suppose that there are in nature superfluous things which do not make sense, namely complete spheres for motions for which a small part of these spheres would suffice . . .' [129]. Why suppose that there are massive spheres to which the planet is fixed when it is sufficient to assume a segment of these spheres produced by two parallel cuts on both sides of the circle along which the planet is carried around in its epicycle? One is thus left with a tambourine to whose rim the planet is attached. The tambourine is rotated in the hollow space between two concentric spherical surfaces which according to the requirements are supposed to be concentric or eccentric to the earth. But of these hollow spheres only the parts enclosing the tambourine need be conceived as real; they are thus two rings or whorls, as Ptolemy characteristically calls them, reminding the reader of Plato's whorls in the *Republic* (616d).

This is all that is left of the system of concentric spheres in Ptolemy's final version, indeed a real 'economy edition'. He again stresses the absurdity of any attempt to combine all these systems of truncated spheres into one by some device similar to the counteracting spheres of Aristotle: 'And senseless, too, are the counteracting spheres, not to mention the enormous increase in numbers which they bring about. For they occupy much space in the aether and are not necessary for the explanation of the planetary motions. They roll back together in one direction in order to produce a single unitary motion . . .' [130].

The principle which secured the unity of the Aristotelian model was a mechanical one, a series of fixed axes whose poles were attached to the enclosing sphere, until the outermost sphere of the fixed stars was reached. What had Ptolemy to offer in place of this mechanism so forcefully discarded by him? The conception of the unmoved movers, of Aristotle's intelligences which kept the spheres in motion, was obviously not to his taste because of its association with the interdependence of all the spheres. The idea of some other force, conceived mathematically or otherwise, had to wait until the establishment of the heliocentric system and the advent of classical mechanics. There remained for Ptolemy, in the age of the return to Pythagoras and Plato, the vitalistic hypothesis of a soul as the driving force of each planet, the soul residing in the planet, and the system of bodies connected with the planet being kept in motion by the vital force emanating from it. Ptolemy illustrates his idea with a simile: 'As an illustration of the motions of the celestial bodies, let us choose birds whose movements are well known to us—we are familiar with illustrations of this kind. The origin of their movements is in their vital force which produces an impulse that spreads into the muscles and from there into their feet or wings, where it ends. . . . There is no cogent reason whatsoever to assume that the motions of all these birds occur through their mutual contact. On the contrary, one has to postulate that they do not touch one another in order that they should not be a mutual hindrance to each other. Similarly, we have to suppose that among the celestial bodies each planet possesses for itself a vital force and moves itself and imparts motion to the bodies united with it by nature' [131]. Ptolemy then explains that the impulses imparted by the vital force to the various parts of the system—the epicycle, the eccentric cycle and so on—need not necessarily be of the same intensity, just as in the human body the impulse emanating from the mind differs in kind and in force from that imparted to the muscles and this again differs from the force of the moving feet—each of these parts having its own natural pull.

Ptolemy thus appears here as a vitalist who attempts to transfer some basic concepts of vitalism to the dynamics of the heavens. Apart from earlier influences which we shall recognize still more clearly in other passages, Ptolemy may possibly have been influenced by the writings of Galen which were widely read and may have come to his knowledge; Galen applies the concept of vital force (*psychikē dynamis*) to the dynamics of animal limbs and, in his work on muscular motion, describes how the vital force regulates the concerted action of the muscles which are thus prevented from obstructing each other and the action of the whole limb by lack of co-ordination.[57] Ptolemy of course insisted on uniform and circular motion as the basic element of celestial kinetics, but the picture of the vital force as the cause of planetary motions was very well suited to his intricate system built up from a multiplicity of complex secondary movements. A return to the mechanistic unity of the Aristotelian picture was out of the question; that picture had been conceived at a time when astronomical knowledge was in a much more primitive state. The simile of a flock of birds, each of them driven by a vital impulse and all of them somehow united and co-ordinated in their flight through the common medium of the air was for him a perfect illustration of the planets and their individual mechanisms carrying out a concerted motion in the aethereal region. In the following passage, Ptolemy combines this picture with the magnificent simile from the *Epinomis*[58] where the celestial motions are compared to a divine dance:

'The parts of the planetary orbits are free to undergo translations and rotations in their natural positions in various ways, except that their movement is uniform revolution, like the chain of hands joined in a circle in a dance, or like the circle of men in a tournament who assist each other and join forces without colliding so as not to be a mutual hindrance. One may illustrate our theory and make it plausible by the construction of an apparatus which explains the eccentric and epicyclic motions. But should somebody use poles when explaining these motions, and insist on their being fixed, he will not arrive at an understanding of the principle of the whole thing nor of its arrangement and the way it works' [132].

These passages will suffice to give us a glimpse of the mentality of a great scientist in the second century A.D. Like all great scientists until today, Ptolemy did not content himself with the material progress of scientific knowledge to which he himself had contributed in no small measure. He felt a deep need for the satisfaction of his epistemological and metaphysical doubts which were raised by the disruptive process of an ever-increasing precision in the description of the geometrical detail. His craving for a physical picture that could restore his belief in some sort of reality, in spite of his positivistic and pragmatic tendencies, is quite understandable, and we may assume that his uneasiness was alleviated by a conception that was in line with the philosophical trends of his time.

3. *Proclus and the Ptolemaic system*

One would indeed think that the conception of Ptolemy with its Platonic leanings should have been welcomed by the Neo-Platonists who flourished for three centuries, from the third till the sixth century A.D. In fact, their reaction was mixed; their admiration was subdued by many reservations and qualifications. This is evident from the writings of such a Neo-Platonic authority as Proclus (fifth century A.D.), especially in his commentary on the *Timaeus* and his *Outline of the Astronomical Hypotheses*. The introduction to the latter book leaves no doubt about the attitude of the orthodox adherents of Plato: they maintained the spirit of what is said in the seventh book of the *Republic* about the limitations of the astronomers and the scientists in general in the face of the true reality. True reality is the unseen reality that can never be grasped by material observation and the experimental procedure of science, but by the light of pure reason and intelligence alone.[59]

Among his introductory remarks, Proclus intersperses some of Plato's phraseology taken verbatim from the *Republic*: 'The great Plato, my friend, expects the true philosopher to take his mind from the perceptible and the totality of changing matter and to transfer astronomy beyond the heavens, to behold there absolute slowness and absolute speed with their true values. From these marvellous sights you seem to lead us down to those orbits in the heaven and to the observations of those practical people, the astronomers, and to those hypotheses which they have artificially devised on the grounds of their observations and which people like Aristarchus, Hipparchus, Ptolemy and others of their calibre used to din into our ears' [124]. Proclus continues that one has to do justice to the labours of these men, in spite of Plato's exhortations, and to give an exposition of their doctrine which they took such great pains to arrive at and to verify. However, he is sure that 'this outline will bring home to you the refutation of the hypotheses on which those people pride themselves and on the grounds of which they develop the whole theory proposed by them' [125].

Proclus regards the Ptolemaic system, of which he gives an account containing several mistakes and inaccuracies, as an ingenious device invented for practical purposes. He takes a decidedly positivistic view and rejects the idea that any reality can be attributed to Ptolemy's spheres or segments of spheres; no mention at all is made of Ptolemy's simile of the birds. A few passages from his commentary on the *Timaeus* follow here: 'These hypotheses have no plausibility at all; on the contrary— some of those invented by later scientists lack the simplicity of divine things, and some suppose the motion of the heavens as if driven by a machine' [119]. The increasing precision of description could cause people to think that Plato's idea was more nearly realized, whereas in fact it was as remote as ever: 'If some people have used epicycles or eccentric circles and assumed uniform motions in order to find numerical values of these motions by a combination of all of them—the epicycles, the eccentrics and the planetary motions on them—one may call this a beautiful invention suited to logical minds which, however, fail to grasp the nature of the whole which only Plato has understood' [120].

Proclus stresses the fact that nowhere does Plato introduce epicycles or eccentrics; further Proclus does not refrain from using Aristotelian arguments to express his disbelief in the reality of Ptolemy's spheres. Simple motions are either in a straight line or in circles, but the circles must have their centres at the centre of the universe: 'It is ridiculous to construct either little circles for each sphere which move in the opposite direction while being parts of the sphere or of some other structure, or eccentric spheres which embrace the centre around which they do not move. These assumptions make void the general doctrine of physics, that every simple motion proceeds either around the centre of the universe or away from or towards it. The epicyclic and eccentric doctrine divides the spheres into segments moving in the opposite direction and it does away with the continuity of each of them, or it introduces circles of a nature foreign to that of the heavens and combines motions out of elements dissimilar and alien to each other because of the dissimilarity of their structures. In view of this, one must bear in mind that Plato never considers the planets as moving in a variety of different ways, because he has no use for mechanical devices that are unworthy of the divine essence. Rather, it is necessary to attribute the diversity to the motion of those souls according to whose wishes the bodies move faster or slower' [121].

Thus we return in the last instance to Plato's world soul and planetary souls as the only reality behind the planetary motions. On the other hand, Proclus declares on several occasions that the epicyclic and eccentric hypotheses are 'not superfluous', because they are a convenient method of resolving complex motions into simple ones, and he recommends the study of this method. However, he cautions his reader that the astronomical experts themselves differ in respect to these hypotheses and also that at times they interpreted their observational results differently. We have to add to this that Proclus himself is not always correct when rendering the theories of the expert astronomers. True, he recounts at great length the intricacies of Ptolemy's system, including the periodic oscillatory changes that occur in the inclinations of the planes of the epicycles of the inner planets to those of the main circles. Astronomy in these centuries was increasingly neglected; a specific instance of its decline is the incorrectness of the treatment of the precession of the equinoxes, Hipparchus' greatest discovery. Ptolemy's computation gave a much poorer result than that of Hipparchus which was indeed very close to the true value. Geminus, Cleomedes and Theon of Smyrna do not mention this important phenomenon at all. Proclus mentions it several times in his *Commentary* as well as in his *Outline*, but he expresses grave doubts of the reality of the phenomenon.[60] Not even the fixed stars are left in peace, he exclaims, but are supposed to change their distance from the pole of the universe. He does not accept this fact and believes that the insistence on it reflects on the astronomers themselves and consequently also on their reliability with regard to other hypotheses. Would not circular motion of the ecliptic around the pole lead to the disappearance of the Great Bear below the horizon? This, he believes, can never happen and is thus evidence against the precession. In another passage Proclus alludes to a curious theory which says that precession is actually not a rotatory but an oscillatory phenomenon.

At the end of his *Outline* Proclus again confesses his doubts and his longing for a 'real' explanation of the celestial phenomena. It is a memorable passage for several reasons, and well worth being quoted in full: 'I shall conclude my book by adding the following to what I have said already. The astronomers who are eager to prove the uniformity of celestial motions are in danger of unconsciously proving that their nature is irregular and full of changes. What shall we say of the eccentrics and the epicycles of which they continually talk? Are these only inventions or have they a real existence in the spheres to which they are fixed? If they are only inventions, their authors have, all unaware, deviated from physical bodies into mathematical concepts and have derived the causes of physical motions from things that do not exist in nature. . . . But if the circles really exist, the astronomers destroy their connection with the spheres to which the circles belong. For they attribute separate motions to the circles and to the spheres and, moreover, motions that, as regards the circles, are not at all equal but in the opposite direction. They confound their mutual distances and sometimes let them coincide in one plane, sometimes separate them and let them cross each other.

Thus there will result all sorts of divisions, foldings-up and separations of the celestial bodies.

'Further, the account given of these mechanical hypotheses seems to be haphazard. Why, in each hypothesis, is the eccentric in this particular state—either fixed or mobile—and the epicycle in that, and the planet moving either in a retrograde or in a direct sense? What are the reasons for those planes and their separations—I mean the *real* reasons that, once understood, will relieve the mind of all its anguish?—this they never tell us. In fact, they proceed in a reverse order: they do not draw conclusions from their hypotheses like the other sciences, but instead they attempt to construct hypotheses which fit those conclusions which should follow from them. . . . However, one should bear in mind that these hypotheses are the most simple and most fitting ones for the divine bodies. They were invented in order to discover the mode of the planetary motions which in reality are as they appear to us, and in order to make the measure inherent in these motions apprehendable' [126].

Proclus concludes on a conciliatory note, but this does not weaken the impression of the whole passage which is a moving revelation of the vacillations of a harassed mind. Is there after all some inner connection between the devices worked out for the sake of convenience and a reality accessible to our mind? And of no lesser interest is the question he raises about the relation between hypothesis and conclusion, a question that made sense in the situation of astronomy after Ptolemy. For the basic hypotheses had not changed—with one short-lived exception—since the days of Eudoxus: taking for granted the geocentric assumption, the celestial phenomena had to be accounted for by allowing only circular and uniform motions. This was relatively simple in the earlier days, but it led to increasing complications as a result of the steadily increasing precision of harmonic analysis in consequence of which Apollonius, Hipparchus and Ptolemy, each in his turn, had to assume new eccentrics or differently moving epicycles or similar changes—in short, 'to construct their hypotheses so as to fit the conclusions'. It is not difficult to guess that what Proclus wants is a 'made to measure' theory that would fit every future observation. When he complains that the astronomers 'do not draw conclusions from their hypotheses like the other sciences', he obviously has in mind the mathematical sciences,

L
149

geometry in particular. However, Proclus is apparently not aware of the fact that if the geometrical conclusions fit the hypotheses, they do so because of the internal consistency of a properly chosen set of axioms, and as a consequence of logical necessity, whereas there is no such necessity and pre-established harmony that can lead from a physical theory to the empirical data. Thus his verdict that the astronomers 'proceed in a reverse order' is not justified. It is an entirely legitimate procedure to modify or extend an hypothesis that has already shown its usefulness in order to make it fit an enlarged or improved set of data. There is of course a limit to such successive and auxiliary *ad hoc* hypotheses, a limit which is not defined by hard-and-fast rules but by common sense. If the process of readjusting the hypotheses loses the character of a correction and the additional assumptions increase out of proportion to the scope of the original ones, then there are good reasons to suspect that the whole theory has to be discarded, to be replaced by a better one, starting with entirely different presuppositions and a new set of hypotheses.

Whether Proclus was or was not motivated by thoughts of this kind when he wrote his concluding remarks, we can but agree with them. The geocentric system had outlived its span of life and, as a purely descriptive theory, had ceased to be adequate to the advanced empirical state of affairs. Looking backwards from our present vantage point we see that only the heliocentric system could offer the start for a new and more fruitful development. Indeed, the purely descriptive phase of this system, associated with the names of Copernicus, Brahe and Kepler, gave way within a relatively short time, with the rise of classical mechanics, to the explanatory phase, beginning with Newton. From then on, the mathematical theories of physics undertake not only to 'save the phenomena' but also to explain them in the framework of systems based on the notion of a physical cause and expressed in well-defined mathematical language. The causal theories, if they do not quite 'relieve the mind of all its anguish', at least satisfy to some extent the epistemological needs of those minds which expect more of a scientific theory than mere material success.

Celestial physics became the subject of a discussion of fundamental significance in the writings of the last great thinker of antiquity, John Philoponus, whose original ideas restored scientific thought to its classical heights. This aspect will be treated in the last chapter of this book, but here we have to refer to Philoponus in another context. The first centuries A.D., until the middle of the sixth, were characterized by an intensive struggle between conflicting and competing spiritual systems of thought. There were strong tendencies within the Christian Church to get rid of the whole pagan *Weltanschauung*, lock, stock, and barrel, including the scientific attitude associated with Greek antiquity.

It was quite natural that celestial physics became a particularly intense focus of this attack. One of the main driving forces behind ancient astronomy had been star worship in an undisguised or sublimated form, and the belief in the divine nature of the stars received fresh support from astrology, imported from the Orient. While both star worship and astrology were vehemently condemned by the Church, there were certain beliefs established within the framework of the monotheistic religions that served as substitutes for this discarded custom and thus preserved its function of upholding the barrier between heaven and earth. An important factor in this respect was the localization of the seat of God and his angels in the heavenly regions. The angels were very useful competition to the spiritual agencies of the pagan world, and it is no wonder that they also replaced the Platonic souls that moved the stars and Ptolemy's vital forces emanating from the planets and driving the spheres.

This doctrine was propounded by Theodorus, Bishop of Mopsovestia, a town in Cilicia, who lived a few decades before Proclus. It was heavily attacked by Philoponus who in his later years attempted to mediate between Greek cosmology and Christian beliefs. *De opificio mundi*, perhaps the last of his works, is an exegesis of the cosmogony of the Scriptures, written with a view to reconciling the biblical story of the creation of the world with the writings of Plato and Aristotle. In his polemics against Theodorus, he advances once again his idea of the impetus, this time as applied to the celestial bodies: 'The supporters of Theodorus' doctrine should tell us where in the Holy Scriptures they have learned that the moon and the sun and each of the planets are moved by angels who either pull them like beasts of draught, or push them, or both together, like people rolling loads in a circle, or else bear them on their shoulders, which would be even more ridiculous? As if God who created the moon, the sun and the other stars could not have invested them with a motive force which is also the inherent cause of the pull of heavy and light bodies and of all the motions originating from the souls of living beings. There is no reason why the angels should force them into motion; for all things that do not move naturally have a forced motion which is contrary to nature and bound to come to an end. How then could the motion of so many great bodies last so long if they were pulled by force?' [100].

Philoponus' *De opificio mundi*, in spite of many clever repartees and ingenious interpretations, shows on the whole the deteriorating effect that a conformist attitude to the Church had on the spirit of scientific inquiry. The Ptolemaic system and all the geometrical descriptions of the celestial motions are dismissed as hypotheses 'entirely divorced from reality'. Nobody, he says, has ever succeeded, nor will ever succeed, in giving a proof of these hypotheses even if he constructs thousands of devices. In this book, reality is even more remote from any conceivable scientific effort than in the Neo-Platonic writings. There is a subdued tone of resignation and a humble attitude of *ignoramus ignorabimus* which is astonishing in view of Philoponus' earlier works written in the spirit of a revolutionary monotheism. One should not inquire into the last causes nor ask too many questions, he now says in a passage which is of interest because in it Philoponus also mentions that the precession of the equinoxes is something inexplicable.

'What is the reason for the number of the spheres, being so and so many according to the old hypotheses and the new ones, and why are there not more or less? Could anybody do the impossible and prove that there must be exactly as many as there are and what the different velocities of the various planets mean . . ., not to mention the motion described by Ptolemy that covers one degree in a hundred years and thus traverses one sign of the zodiac in 3000 years. Who can tell us the cause of these? Nor will anybody ever be able to explain the number of the stars, their position and order and the reason for their various colours. Only this we all believe—that God had created everything beautifully and as it was needed, neither more nor less. Altogether we know the causes of only a few things, and if people cannot tell the natural causes of the manifest things, they should not ask us about the causes of the hidden ones' [101].

This is quite a different Philoponus from the one we have come to know and still more so from him whose acquaintance we shall make in the next chapter. As far as celestial physics is concerned, the spirit of the early Church, for reasons already mentioned, was even more anti-scientific than that manifested in certain tendencies of the late Neo-Platonists. The latter did not allow the light of classical science to be completely extinguished and attempted to find solutions to some of the basic questions that are of great relevance to science.

VI

THE UNITY OF HEAVEN
AND EARTH

---

1. *John Philoponus and his conception of the universe*

IN the previous chapters John Philoponus has been mentioned on several occasions and passages from his works have been quoted. We have come to know his theory of the impetus, his interpretation of light as kinetic energy emitted from luminous bodies and moving according to the laws of geometrical optics, his elaboration of the concept of fitness, and his ingenious extension of the concept of a process 'according to nature' by establishing a theory of perturbation. These instances, among others, have already given ample evidence of the high originality and independence of mind of one of the most remarkable personalities who lived at the very close of antiquity. The present chapter will deal with another contribution by Philoponus to scientific thinking, probably his greatest, seen from the viewpoint of the history of ideas. It concerns his complete negation of the Aristotelian dichotomy between heaven and earth, his radical conception of the universe as a physical unity, an idea which stemmed from his monotheistic convictions as a Christian.

Biographically, little is known about Philoponus[61] or, indeed, about other important personalities of his time, like Damascius or Simplicius. The chronological data of two of his books are given by himself, and thus we know that his commentary on Aristotle's *Physica* was written in A.D. 517 and his book against Proclus in A.D. 529. By indirect inference, through the mention of certain contemporaries, one can conclude that one of his last books, *De opificio mundi*, was written in the middle of the sixth century. Philoponus was thus probably born in the last decades of the fifth century and died in the later half of the sixth. From a remark in his commentary on the *Meteorologica* we know that he was a pupil of Ammonius, one of the last important philosophers of the Alexandrian school who was most probably succeeded in his office by Philoponus. The question of whether Philoponus was born a Christian or was a convert to Christianity during his active philosophical period is still an open one, and inconclusive arguments have been adduced in favour of both assumptions.[62] This problem need not concern us here; what matters for the story told in this chapter is that we have clear evidence of Philoponus' Christian beliefs in some of his extant books and in Simplicius' writings. While he may or may not have written his earlier commentaries on Aristotle as a Christian, there are unmistakable expressions of his monotheistic convictions in his commentary on the *Meteorologica*, in his voluminous book against Proclus, entitled *De aeternitate mundi* (On the Eternity of the Cosmos), and in his exegesis on the biblical story of the creation of the world (already quoted earlier under its Latin title *De opificio mundi*). Philoponus' book against Aristotle (which like the one against Proclus is also called *On the Eternity of the Cosmos*), a book of the greatest relevance in our context, is unfortunately lost, but lengthy excerpts from it have come to us through extensive quotations by Simplicius who vehemently polemized against Philoponus in his commentaries on the *De caelo* and the *Physica*.

Simplicius' polemic against Philoponus, full of bitterness and abuse, has to be judged in the light of the personal histories of these two contemporaries. Whereas the Christian Philoponus' career in Alexandria seems to have taken a smooth and regular course, the life of the Neo-Platonist Simplicius was less easy. The Academy in Athens, where he taught philosophy, was closed in A.D. 529 by the Emperor Justinian; Simplicius with some of his colleagues had to leave, and he spent a number of disappointing years in Persia. Through the vacillations of the uneasy coexistence of Christianity and paganism, these men were able to return to Athens and to continue their work there, if not in official positions, at least privately and apparently unmolested. However, the resentment of Simplicius against his Christian colleague (whom, as he tells us, he never met) can probably be explained as a consequence of the uprooted life of the heathen philosopher during those years. No mention is made of Simplicius in Philoponus' extant writings.

Apart from their different religious beliefs, the picture of the two men, as revealed in their literary works, exhibits other striking contrasts. Simplicius is by far the less imaginative, but of a more ordered mind, very conscientious in referring to his sources in the extensive quotations he gives of earlier writers from pre-Socratic times to his own contemporaries; in this respect the history of philosophy owes him an invaluable debt. As a commentator on Aristotle he is a model of correctness, always at pains not to let his personal views interfere with the interpretation of the text. He does not by any means refrain from stating the diverging Neo-Platonic opinion whenever he sees fit, but he does so after having exhausted all possible representations of Aristotle's ideas and he clearly indicates the critical remarks as digressions from the regular course of his exegesis.

Philoponus, far from being organized, has a rather erratic type of mind, sometimes contradicting himself, easily given to mental associations which divert him from his main theme, and without much regard for reference to sources. His style, too, often reflects a certain lack of discipline and sometimes exceeds in repetitiveness even the usual unwieldy language of his time. However, among all this cumbersome verbosity one comes unexpectedly, again and again, upon passages of great brilliance, ideas sparkling with originality and ingenuity and acute and independent criticism. Especially striking is Philoponus' scientific imagination, his often admirable faculty of giving a concrete illustration of an abstract concept or replacing a worn-out analogy by a new and surprising one.

Although we possess only Simplicius' counter-arguments against Philoponus, we may well look at this series of argument and counter-argument as one of the great dialogues in the history of ideas, as a controversy comparable to Galileo's disputation with his opponents in Florence and Rome or to some of the celebrated exchanges of letters between scientists or philosophers debating the great issues of their period. What was at stake in the discussion between Philoponus and Simplicius was even more than in the great argument that went on in the century before, when

Proclus arose to defend Plato's theory of matter against Aristotle, a defence so aptly seconded later by the same Simplicius. For now it was not a question of method or of scientific approach to a problem. The point at issue was the alternative between two fundamentally different cosmological outlooks, the dominant view of Greek antiquity postulating a dichotomy between heaven and earth, and the view, commonly accepted only from the seventeenth century on, of a physical identity of celestial and terrestrial phenomena.

The unique position of Philoponus in the history of scientific ideas is given by the fact that through him a confrontation of scientific cosmology and monotheism took place for the first time. The very idea on which all monotheistic religions are based implies of course the belief in the universe as a creation of God, and the subsequent assumption that there is no essential difference between things in heaven and on earth. This, as well as the strongest repudiation of star worship, is already clearly expressed in the Bible as the basis of Jewish monotheism and was taken over by the Christian Church and afterwards by Islam. However, neither in classical Hebrew literature nor in the Christian writings preceding Philoponus is any scientific conclusion drawn from these basic tenets of monotheism. The unity of heaven and earth, the sun, moon and the stars being objects created by God just like the grass, the trees, water and the animals—all this was accepted as a fact, it was registered without being interpreted in the frame of a scientific conception or explained in the light of a view of the world differing from former mythological or pagan beliefs.

Such a scientific interpretation was precisely the substance of John Philoponus' argumentation, and his point of departure was the Aristotelian and Neo-Platonic dogma of the eternity of the universe and the invariability of the structure of heaven. For this attack he was well equipped with a thorough knowledge of Aristotle, his Neo-Platonic background acquired in the school of Alexandria, and by being well versed in astronomy and the other branches of Greek science. When he undertook to overthrow the barriers between heaven and earth, Philoponus did not advance any new scientific facts in addition to the established body of knowledge. His was in fact a Copernican position—he saw the universe in the light of a new conception and reinterpreted the known facts accordingly.

2. *Philoponus' arguments and Simplicius' objections*

Philoponus' attack was a concentric one; he attempted to discard every aspect of the dogma of dichotomy and to do so by physical and dialectic arguments, by metaphysical and logical reasoning. He tried to uncover contradictions in Aristotle's system, some-times—to the great fury of Simplicius—with some measure of unfairness; he did not even shrink from sophistries and from con-tradicting himself on occasion. Being convinced of the utter falsity of Aristotle's position he did not take a great deal of care over the consistency of his own; what alone mattered was to prove him wrong from every conceivable point of view. One must see this sometimes daring and not always solid reasoning of Philoponus in the right perspective—the reasoning of a man carried away by his revolutionary zeal and by the momentum of a new and irresistible conception.

One line of attack is Philoponus' attempt to disprove the exis-tence of a fifth imperishable element in heaven, and his most impressive arguments are doubtless the physical proofs he adduces in favour of the fiery nature of the sun and the stars. In the *De caelo* as well as in the *Meteorologica* Aristotle had claimed that 'the stars are neither made of fire nor move in fire' [23], and that the celestial stuff, the aether, 'is eternal, suffers neither growth nor diminution, but is ageless, unalterable and impassive' [21]. Heat and light emitted from the celestial bodies are produced by the friction caused by their movement, similar to the case of flying projectiles: 'Here too the air in the neighbourhood of a projectile becomes hottest' [32]. This is what makes us think that the sun itself possesses the quality of fire, but, adds Aristotle, even the colour of the sun does not suggest a fiery constitution: 'The sun which appears to be the hottest body is white rather than fiery in appearance' [33].

When commenting on the last passage, Philoponus first cor-rects Aristotle with regard to his observation on the colour of the sun and in this connection emphasizes the fact that the colour of a fire depends on the nature of the fuel: 'The sun is not white, of the kind of colour which many stars possess; it obviously appears yellow, like the colour of a flame produced by dry and finely chopped wood. However, even if the sun were white this would not prove that it is not of fire, for the colour of fire changes with the nature of the fuel. Shooting stars and lightnings are white like the colour of the stars (they are also called stars), while the poet called lightning "shining white". The comets too are white, and they obviously are also of fire. The sun itself looks yellow, and even red, when near the horizon. Thus from the colour of the sun it does not necessarily follow that it is not of fire' [89].

The idea that the composition of matter determines the colour of fire (in principle the basis of modern spectroscopy) is repeated in a passage of Philoponus' last work *De opificio mundi*: ' "One star differeth from another star in glory" says Paulus. Indeed, there is much difference among them in magnitude, colour and brightness, and I think that the reason for this is to be found in nothing else than the composition of the matter of which the stars are constituted. They cannot be simple bodies, for how could they differ but for their different constitution? This also causes a very great variation in sublunar fires, in thunderbolts, comets, meteors, shooting stars and lightning. Each of these fires is produced in principle when more or less dense matter is penetrated and inflamed. Terrestrial fires lit for human purposes also differ according to the fuel, be it oil or pitch, reed, papyrus or different kinds of wood, either in a humid or in a dry state' [102].

There could be no more striking example of the complete abandonment of the belief in the divinity of the stars than the equanimity with which Philoponus compares celestial and terrestrial sources of light. In his commentary on the *De caelo* he goes even further, in drawing comparison with luminescent matter: 'And the colour called radiant, and light and all the properties which are attributes of light are also found in many terrestrial bodies, in fire and glow-worms and in the heads or scales of certain fishes and similar objects' [192]. This 'blasphemy' goes a bit too far for the taste of the heathen Simplicius: 'Had Aristotle not given the same name to the celestial light and brightness and to the terrestrial phenomena which are something different, Philoponus would not have dared to say that the light of heaven is also present in glow-worms and in the scales of fishes. . . . But Philoponus takes the light above and the transparent and bright objects here in the same sense, because of his boorish rashness, and declares both to be of the same nature. But why do I blame his lack of education when, whatever he may be, he is apparently completely out of his wits in assuming that the heavenly light is the same as that of a glow-worm? This conceited and quarrelsome man does not realize that David, whom he venerates so deeply, has handed down to him the opposite view. David did not hold the sublunar sphere and heaven as being of the same nature, as is evident from his words ''the heavens declare the glory of God and the Firmament sheweth his handywork'': he does not talk of glow-worms and scales of fishes. Aristotle's saying "accept one paradox and the rest follows" fits this case very well' [194].

Philoponus was the first to refer to colour in the present context, but, in fairness to earlier Greek science, one should not forget the long history of theories of colour from the Presocratics on. It is worth quoting here the most remarkable statement in this field made by Democritus who even established a relation between colour and temperature: 'Iron and other bodies which are heated are brighter when they contain more fire of higher tenuity, and are redder when they contain little fire in a coarse state. Thus redder bodies are less hot' [243]. Besides this very correct observation (which in modern times was the starting-point for the theory of radiation), there were other, incorrect explanations of colour which, like that of Goethe, derived the origin of colour from the contrast of black and white. Thus, according to Theophrastus, 'white appears crimson when seen through black, as for instance the sun when seen through smoke and mist' [245]. Philoponus also referred to it when he spoke of the red colour of the sun near the horizon; in another passage he explains this phenomenon by 'the mingling of the sun's rays with the humid part of the atmosphere'.

Philoponus did not restrict his reasoning to colours but extended it to transparency: 'There is, generally speaking, nothing in the things in heaven which is not found also in terrestrial bodies. Transparency prevails in heaven, and in air and water, and in glass, and is equally found in certain stones' [191]. From here, he takes a further step of generalization: 'What is visible is in principle also tangible, and tangible things possess tangible qualities— hardness, softness, smoothness, roughness, dryness and humidity, as well as heat and cold which contain all the others' [193]. This clear and outright declaration of the identity of celestial and terrestrial matter again provokes Simplicius to abuse and bitter sarcasm, when he hints at the chiliastic beliefs of the Christians:

'We must further mention that this quarrelsome man assumes heat and cold, dryness and humidity, softness and hardness and the other tangible and perceptible qualities to be in heaven. Thus the question arises, if these qualities in heaven are really in mutual action with those on earth, how can one account for the fact that up to now no visible change appears to have occurred in heaven as a result of the influence of the earth? And even supposing that heaven is not easily affected by things on earth, according to them these are already the last days and the end of the world is to be expected very soon, so that by now there should be some change to be seen in heaven and the celestial motions. However, if the things here are influenced by heaven but not the other way around, how dare we say that heaven and earth are of the same nature and even call Plato in as a witness who is supposed to have said that heaven consists of the four elements? How can Philoponus conclude that the sun is hot from the fact that it heats the things on earth and believe it to be full of fire from the evidence of its colour?' [189].

By the same logic, continues Simplicius, introducing arguments from astrology, one should conclude that Saturn is cold and has the qualities of water, because of its cooling influence. This would be the same nonsense as believing in the hot quality of the sun. However, he goes on to say in the true Neo-Platonic spirit, 'all the celestial bodies with their incorporeal powers direct the bodies on earth towards their specific properties like the soul, when in fear or in meditation, by her incorporeal qualities causes the face to blush or the brows to knit' [190]. Philoponus' assumption, Simplicius concludes, is 'a profanation of heaven and of God, its founder and supporter'.

The objection raised by Simplicius, that no change can be observed in heaven, is from a purely empirical point of view a strong argument against Philoponus' thesis. The Egyptians and Babylonians, he says, are in possession of stellar records dating back to times immemorial. 'In all the time during which scientific observations have been recorded, no report is made of any change in heaven with regard to number, magnitude, or colour of the stars, or of their periodic motions' [195]. Again and again, Simplicius stresses this point, sometimes waxing poetic over it: 'If the heaven was created some 6000 years ago, as Philoponus thinks, and if it is already in its last days, as he further pleases to believe, why does it not show any signs of being past its prime and on its way towards decay? At least one thing one should perceive, namely that all movements are getting slower, if old age is really falling upon heaven. However, neither days nor nights nor hours are getting longer, as is proved by comparing all present activities—agriculture, travel and navigation—with those in former times. The distance covered in a day's journey is still the same, and the oxen plough the same area or even less in a day, and the water-clocks, constructed after the same principle, have still the same intake and output of water per hour as in former times' [175].

Simplicius also returns to the other consequence of the removal of the barrier between heaven and earth, the physical interaction of the two regions for which no empirical evidence whatsoever exists. 'Does Philoponus not realize that all things would be transformed into one another if really heaven and the sublunar sphere were composed of the same matter and held the same forms? I cannot believe that even he with all his foolhardy and inconsiderate talk would maintain that things in heaven and on earth could change one into another. If he declared that he could imagine things above being below, he would certainly be regarded as a drunkard among sober people. If matter were the same everywhere, mutual transformations should have happened already many times, for the forms produced in matter last only for a short time' [198].

Philoponus is not at a loss for an answer to these objections, and lack of empirical evidence embarrasses him no more than it did Copernicus when his opponents pointed to the absence of parallax among the fixed stars. Copernicus' answer was the enormous distance of these stars, and that of Philoponus—the great slowness of decay in certain objects or their greater power of endurance which may have natural reasons or be so by the grace of God: 'The fact that in all past time, heaven seems not to have changed either as a whole or in its parts should not be taken as a proof of its being imperishable and uncreated. There are animals which live longer than others, and parts of the earth, such as mountains, stones and hard metals, are all roughly speaking as old as time, and there is no record of Mount Olympus having a beginning, or growing or diminishing. Moreover, for the survival of mortal beings it is necessary that their principal parts should

162

persist in their own natural state. Thus, as long as God wants the universe to exist its principal parts have to be preserved, and admittedly the heaven as a whole and in its parts is the principal and most essential part of the universe' [200].

For Philoponus the main issue is that only God is vested with omnipotence, with the infinite power to act and not to exhaust himself. All material objects and phenomena in the universe have only a limited potential even if they are durable to the extent of permanence. They may be technically permanent, but in principle they are of limited power and duration. We will return later to the Neo-Platonic concept of omnipotence and Philoponus' polemic against its application to the physical world. But first we must see how he attempts to prove that the apparently infinite power of certain objects is just a consequence of their great bulk which creates the impression of permanence. In the passage quoted by Simplicius, Philoponus takes the oceans as an illustration, but the implication for other large objects is obvious: 'It can be proved that the totality of the elements does not possess infinite power. For experience shows that the smaller a quantity of matter the more rapid its decay, and the greater it is the slower the process. Let us assume that a ladleful of water could last a year, and that each equal quantity of water will last the same time. The existing mass of water being finite it could be measured out by ladles and divided into a finite number of them. As each of these parts will be of finite power the same will hold for the whole which is the sum of all the parts; and similarly with the other elements. It is thus proved that each of the four elements in its totality possesses only finite power' [173]. The argument presented by Philoponus is a sound one: slowness of change is not only a function of certain physical properties, such as hardness, as he pointed out before, but also of bulk. And just as change can more easily be observed in a small quantity of water than in the ocean, so a mountain or a big rock seems to be more permanent than a stone. It only remains to be shown that the same applies to the objects in heaven.

Here Philoponus makes use of the term 'omnipotence', going back to Aristotle's concept of infinite potentiality which he discusses in the last chapter of the eighth book of his *Physica*, and which he takes up again in the *Metaphysica*. An infinite body could possess infinite potentiality, but the objects of the universe, being finite, possess finite potentiality, to a greater or lesser degree. God and the intelligences driving the celestial spheres are eternal and thus, by their nature, are perfectly actual. Several of the Neo-Platonists, in their tendency to harmonize Plato and Aristotle, introduced the term 'omnipotence' as an attribute of the eternal world-soul manifest for instance in the eternal revolutions of the celestial bodies: 'To be continually in circular motion is the mark of omnipotence' [118], says Proclus in his commentary on the *Timaeus*, and he repeats this in several other contexts.[63] Proclus is further anxious to prove that Plato, in spite of his assumption of the genesis of the cosmos, did attribute omnipotence to it, as did Aristotle also, in so far as it perseveres in a state of becoming though not in the static one of being. To the totality of the cosmic occurrences one may thus ascribe omnipotence which in the last instance reflects the eternity of the world-soul.

Philoponus combines this Neo-Platonic train of thought with the Aristotelian relation between form and matter and applies the combination to the heavenly bodies, in order to prove the transience of the celestial objects even if it should be suggested that they are made of aether: 'Let us even concede that heaven and the sublunar sphere are not made of the same stuff and suppose that the so-called fifth element, which they believe to be the primary substratum, is the one and only substance of all celestial things. However, this substance happens to exist in different forms (such as that of the sun, the moon and each of the other stars and of every sphere), which obviously shows that the substance of the celestial bodies is fit to receive every celestial form even if for some superior and transcendent reason it should not receive one of these. If now this celestial substance receives different forms it follows that with respect to the potentiality of its matter none of the objects in heaven is imperishable' [171].

For Philoponus, the celestial bodies are not simple bodies but composite ones, they are differentiated by their individuality, their different sizes, movements and periods of revolution. Ultimately, their being matter and form means that they are composite, like everything in the sublunar sphere: 'Just as we take it for granted that one and the same matter is the basis of the innumerable shapes in the sublunar sphere and is fit to receive every form, as is proved by the change of all forms into one another, so obviously the same matter is disposed by nature to hold all the celestial forms' [197]. It does not matter at all whether the celestial stuff is the fifth essence or, as Philoponus believes, identical with the terrestrial elements. For if one abstracts things from their form and from the specific individuality of their matter there remains pure extension, common to all the objects of the physical world. This is brought out clearly in the following fragments from Philoponus' book against Aristotle: 'If the celestial bodies are composite, and composite things imply decomposition, and things implying decomposition imply decay (for the decomposition of the elements is a decay of the composite, and what implies decay has no omnipotence) it follows that the things in heaven, by their own nature, have no omnipotence. . . . Moreover, those who declare heaven to consist not of the four elements but of the fifth essence do assume it to be a composition of the underlying fifth matter and of the solar or lunar form. However, if one abstracts the forms of all things, there obviously remains their tri-dimensional extension only, in which respect there is no difference between any of the celestial and the terrestrial bodies' [172].

The limited power of the celestial bodies cannot be disproved by their permanence or near-permanence, as Philoponus repeatedly emphasizes: 'According to Plato, they may never perish if they are held together by the will of God, by a bond stronger than their own nature.' But this cannot change the essence of Philoponus' argument and of his Cartesian conclusion that all bodies in the universe are substances whose common attribute is extension. Matter everywhere, whether in the sun and the planets, or in terrestrial objects, is nothing but tri-dimensional extension, and, further anticipating Descartes, Philoponus concludes that matter, being a spatial magnitude, must be infinitely divisible, in heaven as well as on earth:

'If every body is infinitely divisible, and the heavenly objects are bodies, they too, by their specific definition, must be infinitely divisible in so far as they have extension, even if they are not actually divided (in the same sense as matter is in itself formless, although it never appears without form). . . . Therefore the division of the heavenly objects will in principle lead to such a size that nothing will remain of their form, just as we can in principle strip matter of its form. If, however, *qua* bodies they admit

M
165

by definition of a material division they will admit, too, of destruction, for we can theoretically carry into actuality what is potentially their attribute. Thus none of the celestial bodies is by its very nature omnipotent. . . . The point in question is the natural law governing each body and not what happens to it by some transcendent cause. One could concede for instance that the celestial bodies, being held together by the divine will, will not perish; this, however, would not exclude that by their specific nature they are subject to the law of destruction' [174].

3. God and Nature

Relentlessly Philoponus applies to the celestial region all the Aristotelian categories of change and becoming. After form and matter, he introduces the third internal cause of change—privation, which brings him to his final step, the concept of the universe as a creation of God: 'To every physical form which has existence in the underlying matter, there exists generally a privation opposed to it, out of which form it is created and into which it dissolves on perishing. Heaven and the whole universe were created in a physical form and thus also presuppose a privation out of which they were created and into which they perish. Just as man is created from non-man and house from non-house and generally, just as each of the forms of nature and art originate from their privation, so heaven which has also a physical form is created from non-heaven and the universe from the non-universe' [196].

This last sentence alludes to the cornerstone of monotheism, the *creatio ex nihilo*. The dogma of the creation of the universe out of nothing went very much against the grain of the Greek mind. Indeed, the history of science shows that it was opposed to the Greek idea of nature. Democritus' principle of the conservation of the existing has already been quoted in another chapter: 'Nothing can come into being from that which is not nor pass away into that which is not' [44]. The strong conviction that a balance exists within the changing phenomena of the physical world derived from the Greek sense of proportion and belief in harmony. One can go right back to the beginning of Greek philosophy, to the doctrines of the Milesian school, and in their approach to nature one finds the presupposition of a rational principle, the belief in the possibility of a rational explanation of things. The attempt to explain the working of things, to find out how nature ticks, represented a new experience of reality, a new *Weltgefuehl*, which pushed the old mythological associations into the background. The universe as it is, continuing from eternity to eternity, was a marvellous revelation of continuous creation of the existing out of the existing, no matter how this creation was explained, be it by the modification of a primary substance, the reshuffling of atoms, the actualization of that which exists potentially, or any other rational assumption. The gods and irrational powers of mythology were not abolished but were rationalized and absorbed in the scheme of nature. Before the advent of monotheism, they were never considered to be 'above nature', but were associated with nature on an equal footing, not reigning above it but within it.

The turning-point in the relationship between God and nature was the conception of the monotheistic religions which put God above nature by establishing him as the creator of the universe *ex nihilo*. Theologically this made God the supreme and omnipotent power, but it was also of deep significance for the subsequent scientific development. This will become clear presently in the dispute between Philoponus and Simplicius, but it is perhaps worth while to look first at the situation which arose in the seventeenth and eighteenth centuries when scientists and philosophers were still deeply concerned about the relation between science and religion. The rapidly and successfully developing mechanistic picture of the world, resting on the solid foundation of mathematical physics, put many scientists in a somewhat apologetic position. It was felt that religion was in danger of being undermined by science, and even before Laplace the question must have been raised whether God was still 'a necessary hypothesis'. Kant, a man of firm deistic views, in the preface to his cosmogony (published 1755), discusses the question at some length. He asks: 'If the universe in all its order and beauty is but the result of the general laws of motion acting upon matter, if the blind mechanics of the forces of nature develops so magnificently out of chaos and arrives by itself at such perfection—there is no longer any validity in the proof of a divine creator which one derives from looking at the beauty of the universe. Nature is self-sufficient, the divine government becomes unnecessary and

Epicurus comes to life again in the midst of Christianity.' After a lengthy analysis, Kant arrives at the conclusion that 'there exists a God for the very reason that nature even in the state of chaos cannot act but regularly and according to law'. God, the creator of matter, has tied matter to certain laws, and matter, once left to these laws, is no longer free to deviate from them. The laws of mechanics which automatically formed the solar system out of the chaotic conglomeration of matter are clear evidence for a primary and omnipotent cause above nature.

The central idea brought out so clearly by Kant's train of thought is, that there is no contradiction between the dogma of God the creator and the conception of a universe functioning like clockwork. On the contrary, there is an obvious relationship between the theological and the mechanistic assumption. The concept of *creatio ex nihilo* involves also the creation of all the attributes of matter including its rational behaviour, and it is just this that makes possible a scientific explanation of the workings of this universe without the redundant supposition of a permanent intervention of its divine author. The sharp division between science and theology cleared the way for a full autonomy of the scientific world picture to an extent which was never reached in the classical and Hellenistic periods of Greek science. The following quotations will illustrate the first emergence of this issue in the sixth century A.D.

The concept of a monotheistic God above nature is clearly stated in Philoponus' words: 'Nature produces its creations from existing things by having matter and action as its foundation and not being able to be or act outside them; however, there is no need for God also, who transcends all that exists, to create matter and action out of existing things, for thus He would not hold anything more than nature. Indeed, God produces not only the forms of things immediately created by Him but it is our belief that He has created matter itself' [165]. There is further another significant distinction between God and nature which puts God on a higher level: divine creation is instantaneous, while natural processes take time: 'Nature is in need of time and becoming in order to create the natural objects, but God gives existence to things immediately created by Him without time and becoming, i.e. without any process of shaping and formation' [166].

With regard to the natural processes—Philoponus continues to argue—in which composite bodies evolve out of each other by the laws of nature, one can follow the chain of development backwards step by step until one arrives at the first elements which were created by God together with the cosmos: 'Matter may become fire out of another fire preceding it, and this out of yet another one, but this chain will finally lead to a fire which was not produced in this way but by friction or some cause different from fire. Similarly it is not impossible that in all processes of the generation of one thing out of another something analogous could be assumed. All things generated by nature one out of another are equally subject to a beginning of their existence. Each species—and this particularly holds for the primary elements— must have a first member which derived its origin not from a similar or dissimilar one preceding it, but was created by God together with the formation of the universe' [167].

This assumption again throws Simplicius completely out of countenance: 'To argue that the elements are now being gener- ated by nature one out of another but that they were in the beginning created by God—what an extraordinary way of a stupid person to inquire into truth! From where else should this generation one out of another come if not from God? How could anybody with a normal mind possibly conceive of such a strange God who first does not act at all, then in a moment becomes the creator of the elements alone, and then again ceases from acting and hands over to nature the generation of the elements one out of another and the generation of all the rest out of the elements?' [168]. To the heathen Simplicius, the deistic conception of a world which, once it has been created by God, continues to exist automatically by natural law is utterly incomprehensible. As against this he states the view of the ancient world that the eternal God rules the eternal and unchangeable heaven and the totality of changing things in the sublunar sphere which, taken as a whole, are eternal too, 'because each of them is immediately under the influence of God, and generation and destruction are interwoven with them through the eternal causes moving in heaven'. Characteristically, Simplicius ends his refutation of Philoponus by quoting the passage from the *Timaeus* (41d) where the Demiurge addresses the newly created celestial gods, i.e. the stars: 'With regard to this, the Demiurge in Plato has said to the new gods: "For the rest, do you, weaving mortal to immortal, make living beings, bring them to birth, feed them, and cause them to grow; and when they fail, receive them back again"' [169].

Creation in time brings up again the old conundrum of time and eternity. Philoponus briefly disposes of the Aristotelian argument that, as time cannot be conceived without a present now, it must stretch infinitely in both directions from the now, since every now is an end as well as a beginning. This, says Philoponus, is begging the question; it is like arguing that a long line whose ends are not visible has infinite extension because every point divides the line into two parts.[64] He also rejects the argument that the conception of a universe created in time pre-supposes that time was existing already before that singular point of creation. His splendid reply is couched in terms calling to mind Spinoza's idea of time: 'Things above time can neither be apprehended nor expressed by us who use temporal language. For just as God apprehends temporal things in a timeless way, so do we apprehend things above time in a temporal way' [170]. We mortals are forced to think in temporal categories and therefore we can only talk of the creation in time. However, we should be careful in drawing conclusions from linguistic analysis: 'If we say "there was not always time" or "there was no time before creation" or use a similar mode of expression having temporal significance, this does not necessarily imply that the idea of time is presupposed in these words, but it simply means that time is not eternal' [170].

There is no need to enter into a thorough discussion of another set of arguments brought by Philoponus against the Aristotelian dichotomy in which he leans on the authority of Xenarchus whose book against the aether was written six hundred years before Philoponus. We have dealt with Xenarchus' anti-Aristotelian considerations in the fifth chapter of this book. Simplicius remarks mischievously that Philoponus is trying hard to outdo Xenarchus in his attack on Aristotle.[65] As only fragments of the relevant books of both authors are extant it is difficult to say to what extent Philoponus has borrowed from the Peripatetic philosopher of the first century B.C. However, two arguments should be mentioned here, both directed against the Aristotelian doctrine of the uniqueness of the circular motion attributed to the aether, and of both Simplicius states expressly the authorship of

Philoponus. The first passage refers to the fact that Aristotle associated the movement of the aether with the conception of a simple and perfect circular motion, namely that having the earth as its centre. After astronomy has discarded the belief in this simplicity of the planetary motions, are we still to adhere to the belief in the uniqueness of the stuff of which the stars are made?

Philoponus, referring to Aristotle and to the most orthodox of his commentators Alexander, argues thus: 'If Alexander has correctly observed that Aristotle calls motion in the proper sense that which goes round the centre of the universe, those motions which are not round this centre are neither circular in the proper sense nor simple. According to the astronomers, the stars each have their own specific movements along their spheres and around their own centres, not homocentric with the universe. Thus obviously neither these stars nor their epicycles nor the so-called eccentric spheres possess movements which are circular in the proper sense or which are simple, because upward and downward components can be observed in them. All this is contrary to the Aristotelian hypotheses and, moreover, the stars appear quite distinctly sometimes nearer to the earth and sometimes farther away' [186] [66] The essentially equal nature of heaven and earth once being presupposed, the spell of the Aristotelian classifications is completely broken. The celestial motions are complex compared with Aristotle's original conception of circularity; analysis into epicycles restores circularity, but not in the sense which induced Aristotle to introduce the fifth element.

In his other argument, Philoponus turns against Aristotle's view that the uniqueness of circular motions can be demonstrated by its having no opposite. Aristotle pointed out that the terrestrial motions up and down are opposite in the sense that they lead to, and end in, opposite natural places. Circular motions however, be they clockwise or anti-clockwise (to use a modern expression), always arrive back at the same point and are thus essentially of one type. [67]

Philoponus' reply is again a clear proof of the acuteness of his mind. In order to illustrate his point that their different senses of direction constitute clockwise and anti-clockwise motions as opposite movements, he chooses an example taken from the celestial region: 'Though the starting-point and terminal point of circular motions are the same, there is still an antagonism of the opposites because the direction of the start of one of them is the direction in which the other ends. For instance, of two motions starting in the constellation of Aries, the one leads outwards in the westward sense to the Fishes and Aquarius and so on, and the other leads inwards in the eastward sense, to Taurus and Gemini' [202]. Philoponus continues by reminding us that the first sequence gives us the sense of rotation of the sphere of the fixed stars, whereas the second sequence is that of the sense in which the circular motion of the planets takes place. The implication of this simile is obvious: in heaven itself the two so essentially different movements of the daily rotation and the individual revolution of the planets have opposite senses of direction. The clockwise and anti-clockwise sequence of the zodiacal constellations define opposite motions of eminently concrete significance. Again the conclusion is the same as before—there is no justification for the aether, because celestial and terrestrial movements are of the same character in each having its opposite.

But Philoponus does not stop at destroying the theoretical foundations of the conception of the aether; he carries his attack to the sublunar region and questions the distinction between the two natural terrestrial movements, the upward and downward motions. Natural locomotion, he argues, is but one of the many types of natural change that can occur in a body. With respect to other types of change, such as those of colour, volume, temperature, we observe in nature that one and the same body can undergo change in opposite directions—it can become black or white, can grow or diminish in bulk, can get warm or cold. However, he points out, the same is also true for so-called natural locomotion: 'Air does not only possess the principle of upward motion but also that of downward motion. For if some of the earth or the water beneath it is removed, air immediately fills its place and likewise when something above is removed, air moves upward. If one may attribute downward motion to the force of the vacuum and not to a natural principle, what prevents us from saying that upward motion too has the same cause? Air may be moving up if there happens to be an empty space, and otherwise not' [201].[68] The conclusion is that the same element can obviously move in opposite directions, not because of inherent natural motion but because in both cases there exists a natural external cause which produces it. These considerations of Philoponus, which may well have been based on Hero's pneumatics, also had their share in undermining the basic conceptions of Aristotle's dynamics.

We can content ourselves with these few excerpts from that part of Philoponus' anti-Aristotelian writings which may or may not have been influenced by Xenarchus. Together with previous quotations, it gives an adequate picture of a doctrine which leaves intact very little of a system that was dominant for the greater part of antiquity and, later, throughout the Middle Ages. Simplicius' vehement reaction gives an inkling of how these ideas were probably received among Neo-Platonic and other pagan contemporaries of Philoponus. We do not know how his fellow-Christians received them, in Alexandria where he taught, or in the other parts of the Greek-speaking world. Philoponus belonged to the monophysite sect which flourished in his time in Egypt and in Syria; Aristotle's writings were held in high esteem in these countries as can be inferred from the fact that many of his works were translated into Syriac. It may well be that the positive impression of Philoponus' radical monotheism was outweighed by the negative one of his strong anti-Aristotelian bias. Whether he got a hint from the monophysite Patriarch Sergius, or whether he was induced by some other external or internal cause, his attempt to harmonize his monotheistic belief with Aristotle, in his last book *De opificio mundi*, comes as an anticlimax after his writings against Proclus and Aristotle. This commentary on biblical cosmogony is written in the spirit of Patristic homilies like that of Basilius (fourth century A.D.) whom Philoponus quotes frequently. The tendency of the book is to show that the story of the Bible is in accord with contemporary conceptions of nature and with Aristotle and Plato, Hipparchus and Ptolemy. Concurrent with homiletic talk and much exegesis in a some-times subdued spirit the book here and there reveals sparks of Philoponus' former boldness and radicalism. A few significant passages have been quoted earlier, but perhaps still more characteristic of this last work of Philoponus is another passage in which he curiously mixes natural history and interpretation of Scripture. When talking about fire and light (while discussing the difference between the light created on the first day and the celestial lights created on the fourth) Philoponus again mentions terrestrial light sources which do not burn such as glow-worms, scales of certain fishes and bones of certain animals, but he also mentions the flame of the fire which appeared to Moses out of the bush and did not consume it.[69]

That *De opificio mundi* was written after the book against Aristotle does not necessarily mean that Philoponus disowned his earlier ideas, but it is certainly typical of these ideas that they found no positive response at the time when they were made publicly known, although some centuries later they had considerable influence on Arabic thought. The main theme of Philoponus' attack was the essential equality of heaven and earth, a thesis which he attempted to prove by empirical and analytical arguments. By abrogating the superiority of heaven and putting this particular aspect of monotheism in the forefront he probably provoked indignation among the Christians, no less than he hurt pagan feelings. For despite the resolved fight of Christianity against astrology and the belief in the divinity of stars, the preponderance of heaven was still far from being abolished in the consciousness of the monotheistic religions, and persisted for many centuries to come. The old deities, the stars, were dethroned but they were replaced by God and his angels whose seat, too, was localized in heaven. The whole religious topography, reserving heaven for God and his entourage and associating the subterranean region with the abode of the wicked, was not only deeply rooted in popular belief, but had the support of official teachings of the Church and was not considered merely as allegory.

It was thus unfortunate, from a tactical point of view, that Philoponus displayed such an uncompromising attitude in stressing the omnipresence of God in a universe which in its entirety is subject to the laws of change, as is brought out so clearly in, for instance, the following of his fragments: 'If people assign the place above to the Divinity, this is not yet to be taken as a proof that heaven is imperishable. Because those too who believe the holy places and temples to be full of gods and raise their hands towards them do not assume these dwellings to be without beginning or imperishable but regard them only as a place more fit than others to be inhabited by God' [199].

One is tempted to speculate on how the course of the history of ideas would have been changed had the doctrine of Philoponus been accepted by the Church instead of the Aristotelian conceptions. Had for instance Thomas Aquinas chosen Philoponus' ideas and incorporated them in the scientific foundations of Christian philosophy, the birth pangs of the Copernican and Galilean revolution would perhaps have been less severe and scientific progress possibly accelerated. Though there will never be an answer to the question of whether the by-passing of Philoponus' brilliant ideas was one of the 'missed opportunities' in the history of science, one can see many reasons for their being disregarded. Besides the impossibility of succeeding against the tremendous odds of the deep-rooted religious 'topography' there was the fact that the Monophysites were declared heretics in the century following Philoponus. Thus while he lived on in the scholastic tradition under the modest label of an Aristotelian commentator, his original ideas were far less acceptable to the Church than the allegories of the pagan Plato or the scientific philosophy of the pagan Aristotle.

Philoponus' unsystematic and sometimes abrupt manner of expression and his clumsy style were also a serious handicap to the promulgation of his ideas. But by far the most powerful reason for the non-acceptance of his doctrine was that it was proclaimed at the very end of antiquity when scientific thinking was fading away and the few centres of learning thinning out. Traditional school routine could still continue, but the mental environment necessary for the development of a new and revolutionary scientific conception no longer existed.

The student of the last centuries of antiquity is left with a feeling of frustration and perplexity. Looking back at that period one can see two distinct signs of a renaissance of scientific thought. Iamblichus, Proclus and Simplicius clearly recognized the significance of mathematical quantities as symbols for the description of nature, and Philoponus anticipated the breakdown of the barrier between heaven and earth. In the former case, lack of an algebraic notation was one of the additional reasons for preventing developments which in modern theoretical physics have so completely vindicated the Neo-Platonists. Moreover, systematic experimentation was still far away, and the productive period of Greek science had come to its close; consequently Philoponus' ideas in their isolation could not survive either, and they were not vindicated until Galileo. Despite the level of scientific sophistication and maturity reached at the close of antiquity the world had to wait for many centuries for the continuation of the story of science.

175

NOTES

1. *The Leibniz-Clarke Correspondence*, ed. H. G. Alexander, New York, 1956, pp. 25–26.
2. Simplicius, *Phys.*, 615, 34.
3. Proclus, *In Tim.*, 164b, 165b, 348b.
4. loc. cit., p. 27.
5. Before Strato, Aristotle has stated in general terms that 'since time is the measure of motion, it will also be, indirectly, the measure of rest; for all rest is in time.' (*Phys.*, 221b 7.)
6. Simplicius, *Categ.*, 348, 2.
7. For a more detailed account of the Stoic conception of time see *Physics of the Stoics*, pp. 98–105.
8. Aristotle, *Phys.*, 223b 22.
9. J. E. McTaggart's article 'The Unreality of Time' appeared in *Mind*, vol. 17 (1908), pp. 457 f. For K. Goedel's cosmological model and his comments on McTaggart, see *A. Einstein: Philosopher-Scientist*, ed. P. A. Schilpp, Evanston, Ill., 1949, p. 557.
10. The subject discussed in this section is dealt with in my paper 'Conceptual Developments in Greek Atomism', *Archives Internationales d'Histoire des Sciences*, vol. 11 (1958), pp. 251 f.
11. The last part of quotation 45, from 'unless' on, has wrongly been considered to be a later interpolation, as P. v. d. Muehll points out in his edition of Epicurus' letters.
12. Aristotle, *De gener. et corr.*, 327a, 15–23.
13. *De part. animal.*, II, 2–3.
14. Anaxagoras' statement on the colour of snow: Cicero, *Acad.*, II, 100, and Sextus Empiricus, *Pyrrh. hypot.*, I, 33.
15. On the beginnings of functional thought, see *Physics of the Stoics*, pp. 81–88.
16. Cf. Porphyr., *Vita Plot.*, 14, 7, and Simpl., *De caelo*, 20, 21; 37, 33.
17. Plato, *Republ.*, 510–511, and Iamblichus, *On the Comm. Math. Science*, ch. 8.
18. Aristotle, *De caelo*, 306a, 25.
19. Cf. Simpl., *De caelo*, 660–661.
20. See also Simpl., *De caelo*, 641, 21–28.
21. *Timaeus*, 18b.
22. *Timaeus*, 43a.
23. Cf. Philoponus, *Phys.*, 639, 3–642, 26. Excerpts are translated in

Cohen-Drabkin, *A Source Book in Greek Science*, New York, 1948, pp. 221 f.

24. Aristotle, *De caelo*, 273b 30, 277b 4, 290a 1, 309b 13.
25. Cf. Simpl., *Phys.*, 916, 4, and Cohen-Drabkin, p. 211.
26. Cf. e.g. Arist., *De caelo*, 277a 27.
27. Arist., *De caelo*, 311b 5–13.
28. Vitruvius, *De architectura*, IX. Cf. also ibid., VII, 8.3 (transl. in Cohen-Drabkin, p. 238).
29. 'Carmen de ponderibus', in *Metrolog. Script. Reliqu.*, ed. F. Hultsch, Lepizig, 1866, pp. 88–98. For another description of the hydrometer in the fifth century see Cohen-Drabkin, p. 240.
30. Arist., *De gener. et corr.*, 329b 18.
31. Arist., *Categ.*, 5b 1.
32. Ibid., 4b 20.
33. Cf. Simpl., *Categ.*, 128, 20–24.
34. Philop., *Phys.*, 420 f.
35. Hero, *Pneumatica*, I, Introduction.
36. Philop., *Phys.*, 309, 29–310, 15.
37. Simpl., *Phys.*, 84, 15–86, 18. Cf. also Themist., *Phys.*, 5, 26 f.
38. Arist., *Meteor.*, 378a 26.
39. Cf. Iamblichus, *Myster.*, 5, 20, and Proclus, *In Parmen.*, 927.
40. Cicero, *De republ.*, I, 14.
41. Cf. E. R. Dodds: Proclus, *The Elements of Theology* (Oxford, 1933), prop. 39 and prop. 140, and pp. 222 and 273 n.
42. Galen, *De natur. facult.*, II, 126.
43. Hero, *Pneumatica*, I, 40.
44. Hero, *Automatopoetica*, pp. 340–342.
45. The subject discussed in the following is dealt with at greater length in my paper 'Philoponus' Interpretation of Aristotle's Theory of Light', *Osiris*, vol. 13 (1958), pp. 114 f.
46. Arist., *Polit.*, 1256b 20.
47. Hero, *Catroptica*, p. 324.
48. Maupertuis' principle is stated in his *Essai de Cosmologie* (1751), pp. 221–222; Fermat's principle (1657): Fermat, *Œuvres* (1891), vol. II, p. 354.
49. Plato, *Ion*, 533d.
50. Galen, *De natur. facult.*, I, 28; 48–55.
51. Alex. Aphr., *Quaest. et solut.*, 74, 4–30. The text is obviously corrupted.
52. Cf. Arist., *Meteor.*, 339a 25, and Philoponus' commentary on this passage, 12, 7.
53. L. Euler, *Lettres à une Princesse d'Allemagne* (1787), vol. I, p. 289.
54. Arist., *Meteor.*, I, ch. 4 and 7.
55. Whereas the divine stars are made of fire, the *Epinomis* (984b f.) also mentions the aether, however, as a stuff of which the 'spirits' (intermediaries between the gods and men) are made.
56. Simpl., *De caelo*, 20, 10–25.

57. Galen, *De muscul. motu*, Kuehn, IV, pp. 403–404.

58. *Epinomis*, 982c.

59. *Republ.*, 529–531.

60. Cf. *Outline*, I, 26; III, 54; VII, 45; and *In Tim.*, 277d–278a.

61. See Gudeman's article on Ioannes Philoponus in Pauly-Wissowa's *Real-Encyclopaedie der classischen Altertumswissenschaft.*

62. Gudeman, loc. cit., and E. Evrard: 'Les convictions réligieuses de Jean Philopon et le date de son Commentaire aux Méteorologiques', *Bull. Acad. roy. de Belgique* (*Classe de Lettres*), VI (1953), pp. 299 f.

63. Omnipotence in Proclus: *In Tim.*, I, 294, 28–295, 12; II, 131, 4 f.; 262, 5 f.; III, 21, 2 f.

64. Cf. Arist., *Phys.*, 251b 20–27, and Philoponus' argument in Simpl., *Phys.*, 1166,37 f.

65. Simpl., *De caelo*, 25, 22–34.

66. P. Duhem, in *Le Système du Monde*, vol. II, p. 61, mistakenly attributes this fragment to Xenarchus.

67. Cf. Arist., *De caelo*, 270b 32–271a 29.

68. Philoponus evidently goes much further than Aristotle in *De caelo*, 312b 6 f.

69. Philop., *De opificio mundi*, 186, 3–22.

SOURCES

Alexander Aphrodisiensis, *Scripta minora*, ed. I. Bruns (Berlin Academy), Berlin, 1887.

Ammonius, *In categ. comment.*, ed. A. Busse (Berlin Academy), Berlin, 1895.

Aristotle, *Anal. poster.*, ed. W. D. Ross, transl. G. R. G. Mure, Oxford, 1928.

— *Physica*, ed. and transl. Ph. H. Wicksteed and F. M. Cornford (Loeb), London, 1957.

— *De caelo*, ed. and transl. W. K. C. Guthrie (Loeb), London, 1953.

— *De gener. et corr.*, ed. and transl. E. S. Forster (Loeb), London, 1955.

— *Meteor.*, ed. and transl. H. D. P. Lee (Loeb), London, 1952.

— *De anima*, ed. and transl. W. S. Hett (Loeb), London, 1957.

— *Metaph.*, ed and transl. H. Tredennick (Loeb), London, 1957.

— *Opera minora*, ed. and transl. W. S. Hett (Loeb), London, 1955.

Diogenes Laertius, *Vitae philos.*, ed. and transl. R. D. Hicks (Loeb), London, 1950.

Epicurus, *Epistulae*, ed. P. v. d. Muehll (Teubner), Lepizig, 1922.

Galenus, *De temperamentis*, ed. G. Helmreich (Teubner), Leipzig, 1904.

Hero Alexandrinus, *Mechanica*, ed. L. Nix and W. Schmidt (Teubner), Leipzig, 1900.

Iamblichus, *De comm. mathem. scient.*, ed. N. Festa (Teubner), Leipzig, 1891.

Ioannes Philoponus, *In phys. comment.*, ed H. Vitelli (Berlin Academy), Berlin, 1887.

— *In de gener. et corr. comment.*, ed. H. Vitelli (Berlin Academy), Berlin, 1897.

— *In meteor. comment.*, ed. M. Hayduck (Berlin Academy), Berlin, 1901.

— *In de anima comment.*, ed. M. Hayduck (Berlin Academy), Berlin, 1897.

— *De opificio mundi*, ed. G. Reichardt (Teubner), Leipzig, 1897.

Olympiodorus, *In meteor. comment.*, ed. A. Busse (Berlin Academy), Berlin, 1902.

Philo Alexandrinus, *Opera*, ed. and transl. F. H. Colson and G. H. Whitaker (Loeb), London, 1949.

Plato, *Opera*, ed. J. Burnet, Oxford, 1937.

179

Plotinus, *Enneades*, ed. and transl. E. Brehier, Paris, 1954.
Plutarch, *De Isid. et Osir.*, ed. and transl. F. C. Babbitt (Loeb), London, 1957.
— *De primo frigido*, ed. and transl. W. C. Helmbold (Loeb), London, 1957.
— *De Stoic. repugn.*, ed. M. Pohlenz (Teubner), Leipzig, 1959.
Proclus Diadochus, *In Tim. comment.*, ed. E. Diehl (Teubner), Leipzig, 1903.
— *In Rem. Publ. comment.*, ed. W. Kroll (Teubner), Leipzig, 1901.
— *Hypotyp. astron. posit.*, ed. C. Manitius (Teubner), Leipzig, 1909.
Claudius Ptolemaeus, *Syntax. mathem.*, ed. J. L. Heiberg (Teubner), Leipzig, 1903.
— *Opera astron. minora*, ed. J. L. Heiberg (Teubner), Leipzig, 1907,
Simplicius, *In categ. comment.*, ed. C. Kalbfleisch (Berlin Academy). Berlin, 1907.
— *In phys. comment.*, ed. H. Diels (Berlin Academy), Berlin, 1882.
— *In de caelo comment.*, ed. J. L. Heiberg (Berlin Academy), Berlin, 1894.
Ioannes Stobaeus, *Eclogae*, ed. A. Meinecke (Teubner), Leipzig, 1860.
Strabo, *Geographica*, ed. A. Meinecke (Teubner), Leipzig, 1913.
Themistius, *In phys. paraphras.*, ed. H. Schenkl (Berlin Academy), Berlin, 1900.
Theo Smyrnaeus, *Expos. rer. mathem.*, ed. E. Hiller (Teubner), Leipzig, 1878.
Theophrastus, *Opera omnia*, ed. F. Wimmer, Paris, 1931.

INDEX OF PASSAGES QUOTED

Available translations, mainly those in the Loeb Classical Library edition, have been used, sometimes with minor changes. Quotation 144 is given in the translation of Th. Heath (*Greek Astronomy*, London 1934). All other texts are in the author's own translation.

GENERAL INDEX

Weight—*contd.*
  natural place, 77ff
  as quantity, 83
  specific, 81f
  as a volume force, 85
Whole and its parts, 67f, 97
World soul, 147, 164

Xenarchus, 124
  his life, 126
  on simple lines and motions,
    127ff
  a source for Philoponus, 130, 170

Zenodorus, 26